A-Level Mathematics

Exam Board: OCR MEI

A-Level Maths exams are a piece of cake... *rock* cake, that is. [*Silence*.] They're hard, is what we're trying to say. So you're going to need plenty of practice to do well.

By incredible coincidence, this CGP book is bursting with exam-style questions for every A-Level topic — there's even a full set of practice papers at the end to make sure you're ready for the real exams. Sounds like just your sort of thing, actually.

What else can we tell you? Well, everything's been crafted to cover both years of the latest OCR MEI course, and we've included step-by-step solutions and full mark schemes at the back. Go on, get your teeth into it.

A-Level revision? It has to be CGP!

Published by CGP

Editors:
Chris Corrall, Sammy El-Bahrawy, Will Garrison, Simon Little, Alison Palin, Andy Park, David Ryan, Caley Simpson

Contributors:
Kevin Bennett, Mark Moody, Kieran Wardell, Charlotte Young

ISBN: 978 1 78294 743 1

With thanks to Janet Dickinson, Shaun Harrogate and Dawn Wright for the proofreading.
With thanks to Emily Smith for the copyright research.

Clipart from Corel®
Printed by Elanders Ltd, Newcastle upon Tyne

Based on the classic CGP style created by Richard Parsons.

Contents

☑ Use the tick boxes to check off the topics you've completed.

Exam Advice

Good exam technique can make a big difference to your mark, so make sure you read this stuff carefully.

Get familiar with the **Exam Structure**

For **A-level Mathematics**, you'll be sitting **three papers**.

For the Comprehension part of Paper 3, you'll be given some information to read on a Pure Maths topic, which you'll then have to answer questions on.

Pure Mathematics and Mechanics **2 hours**	**100** marks	**36.4%** of your A-level	Covers the topics tested in **Section One & Three** of this book.
Pure Mathematics and Statistics **2 hours**	**100** marks	**36.4%** of your A-level	Covers the topics tested in **Section One & Two** of this book.
Pure Mathematics and Comprehension **2 hours**	**75** marks	**27.3%** of your A-level	Covers the topics tested in **Sections One** of this book.

Some *formulas are given in the* ***Question Paper***

The exam paper will contain a **formula sheet** in the front that lists some of the formulas you might need.
The formulas you'll be given are shown on pages 164-166 of this book.
You don't need to learn these formulas but you do need to know **how to use** them.

Manage Your Time *sensibly*

1) The **number of marks** tells you roughly **how long** to spend on a question — you've got just over a minute per mark in the exam. If you get stuck on a question for too long, it may be best to **move on** so you don't run out of time for the others.
2) You don't have to work through the paper **in order** — you could leave questions on topics you find harder until last.

Get **Familiar** with the **Large Data Sets**

Throughout your course you'll be working with up to **three large data sets**. In the exam, **only one** of these will be examined — and this will change each year. The large data set will only be used in Paper 2 for A-level Maths.

We've used **three** sets in the questions in Section Two of this book,
and chosen **one** of the data sets to use for Practice Paper 2.
Make sure you know **which set** will be used in **your** exam, and that you're **comfortable** with that one.

Questions using the large data set might:

- Assume that you're familiar with the **terminology** and **contexts** of the data.
- Use **summary statistics** based on the large data set — this might reduce the time needed for some calculations.
- Include **statistical diagrams** based on the large data set.
- Be based on a **sample** from the large data set.

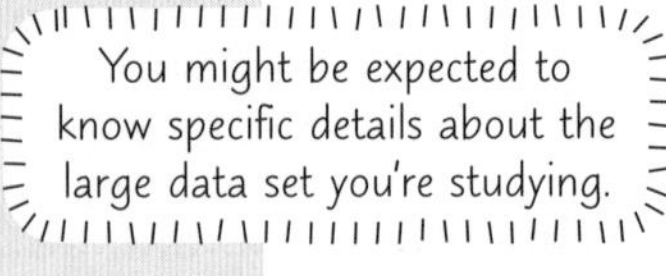

Watch out for ***Modelling*** *and* ***Problem-Solving*** *questions*

The A-level Maths course has a few **overarching themes** — **proof**, **problem-solving** and **modelling**.
The first topic in this book covers proof (and there are other proof questions dotted throughout the book).
Problem-solving and modelling questions are covered throughout the book, but they can be trickier to spot.

- **Problem-solving** questions involve skills such as **combining** different areas of maths or **interpreting information** given to identify what's being asked for. They're often worth a **lot of marks**, as there's a lot of maths involved in them.
- **Modelling** questions involve using maths to represent **real-life situations**. You might be asked to think about the **validity** of the model (how realistic it is) or to interpret values **in context**.

Proof

Welcome one and all to this wondrous (if I do say so myself) A-Level practice book. First up is a real tough cookie — proof. But, if you prove yourself worthy on this section, you'll be well on your way to success...

1 Prove the following statement:

For all integers n, $n^3 - n$ is always even.

(2 marks)

2 Prove that there is no largest integer.

(3 marks)

3 Prove that the product of any two distinct prime numbers has exactly four factors.

(3 marks)

4 Prove that if a and b are rational numbers, the number $c = a - b$ is also rational.

(3 marks)

5 Prove that $n^3 + 2n^2 + 12n$ always has a factor of 8 when n is even.

(3 marks)

Proof

6 A student makes the following statement:

"If x^3 is even, then x is even."

She attempts to prove the claim as follows:
"Let $x = 2k$. Then $x^3 = (2k)^3 = 8k^3 = 2(4k^3)$ which must be even, therefore the claim is true."

a) Explain why the student's proof is not valid.

....

....

....

(1 mark)

b) Prove the student's statement.

(4 marks)

7 Riyad claims that, "if x and y are both irrational, then $\frac{x}{y}$ is also irrational".

a) Disprove Riyad's claim with a counter-example.

(1 mark)

Riyad goes on to claim that "any non-zero rational number multiplied by any irrational number is irrational."

b) Prove Riyad's claim by contradiction.

(3 marks)

When you're doing a proof in an exam, it's really important that it's laid out in a clear and logical manner. If it's a proof by contradiction, state the assumption you've made and then show how this assumption leads to a contradiction. Once you have the contradiction, don't just stop — finish with a concluding line starting "hence" or "therefore".

Score

23

Algebra and Functions 1

Algebra is a pretty important part of maths — so it's a good idea to get to grips with it now. First up, Algebra and Functions 1, where you can practise all things surd-like and fraction-y. The excitement is almost palpable.

1 Simplify $\frac{a^6 \times a^3}{\sqrt{a^4}} \div a^{\frac{1}{2}}$. Give your answer in the form $a^{\frac{p}{q}}$.

....

(2 marks)

2 Find the value of x such that:

a) $27^x = 3$

$x =$ ....

(1 mark)

b) $27^x = 81$

$x =$ ....

(1 mark)

3 Fully simplify $\frac{(3ab^3)^2 \times 2a^6}{6a^4b}$.

....

(2 marks)

4 Show that $\frac{(5+4\sqrt{x})^2}{2x}$ can be written as $\frac{25}{2}x^{-1} + Px^{-\frac{1}{2}} + Q$, where P and Q are integers.

(3 marks)

Algebra and Functions 1

5 Express $(5\sqrt{5} + 2\sqrt{3})^2$ in the form $a + b\sqrt{c}$, where a, b and c are integers to be found.

....

(4 marks)

6 Rationalise the denominator of $\frac{10}{\sqrt{5} + 1}$. Give your answer in its simplest form.

....

(3 marks)

7 Express $\frac{4 + \sqrt{2}}{2 + \sqrt{2}}$ in the form $a + b\sqrt{2}$, where a and b are integers.

....

(3 marks)

8 Express $\frac{(x^2 - 9)(3x^2 - 10x - 8)}{(6x + 4)(x^2 - 7x + 12)}$ as a fraction in its simplest form.

....

(3 marks)

Algebra and Functions 1

9 For this question, give all answers as fractions in their simplest form.

a) Simplify $\frac{x^2 + 5x - 14}{2x^2 - 4x}$.

....

(2 marks)

b) Using your answer to part a) or otherwise, write $\frac{x^2 + 5x - 14}{2x^2 - 4x} + \frac{14}{x(x-4)}$ as a single fraction.

....

(3 marks)

10 Express $\frac{1}{x(2x-3)}$ in partial fractions.

....

(3 marks)

11 Express $\frac{6x-1}{x^2 + 4x + 4}$ in partial fractions.

You'll have to factorise the denominator first.

....

(4 marks)

In an exam, you might not always be asked to write an expression as partial fractions, but if the denominator is a product of two linear factors, it's usually a good place to start. There's so much algebra to cover in A-Level Maths, I've decided to split it up into a few parts. This is the end of Part 1. Next, with crushing inevitability, comes Part 2...

Score

34

Algebra and Functions 2

Algebra and Functions Part 2 will take you on a whistle-stop tour of quadratic equations, make a slight detour into the world of simultaneous equations and inequalities and finish with everyone's favourite — cubics.

1 Given that the equation $3jx - jx^2 + 1 = 0$, where j is a constant, has no real roots, find the range of possible values of j.

....

(3 marks)

2 $f(x) = \frac{1}{x^2 - 7x + 17}$

a) Express $x^2 - 7x + 17$ in the form $(x - m)^2 + n$, where m and n are constants.

....

(3 marks)

b) Hence find the maximum value of $f(x)$.

....

(2 marks)

3 Find the possible values of k if the equation $g(x) = 0$ is to have two distinct real roots, where $g(x)$ is given by $g(x) = 3kx^2 + kx + 2$.

....

(3 marks)

4 Solve the equation $x^6 = 7x^3 + 8$.

....

(4 marks)

Algebra and Functions 2

5 A scientist working at a remote Arctic research station monitors the temperature during the hours of daylight. For a day with 9 hours of sunlight, she models the temperature using the function $T = 10h - h^2 - 27$, where T is the temperature in °C and h is the time in hours since sunrise.

a) **(i)** Express $T = 10h - h^2 - 27$ in the form $T = -(m - h)^2 + n$, where m and n are integers.

....

(3 marks)

(ii) Hence show that T is always negative according to this model.

(1 mark)

b) **(i)** State the maximum temperature predicted by this model, and state the number of hours after sunrise at which it will occur.

Temperature = ... °C, time = ... hours after sunrise

(2 marks)

(ii) Sketch the graph of $T = 10h - h^2 - 27$ on the axes below. Mark clearly on your graph the temperature at sunrise.

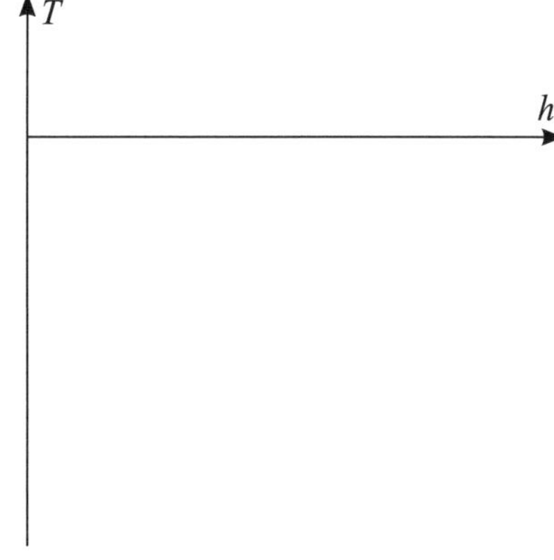

(2 marks)

6 Solve the simultaneous equations $y + x = 7$ and $y = x^2 + 3x - 5$.

x = ... y = ... or x = ... y = ...

(4 marks)

Algebra and Functions 2

7 The curve C has equation $y = -x^2 + 3$ and the line l has equation $y = -2x + 4$.

a) Find the coordinates of the point (or points) of intersection of C and l.

....

(3 marks)

b) Sketch the graphs of C and l on the axes below, clearly showing where the graphs intersect the x- and y- axes.

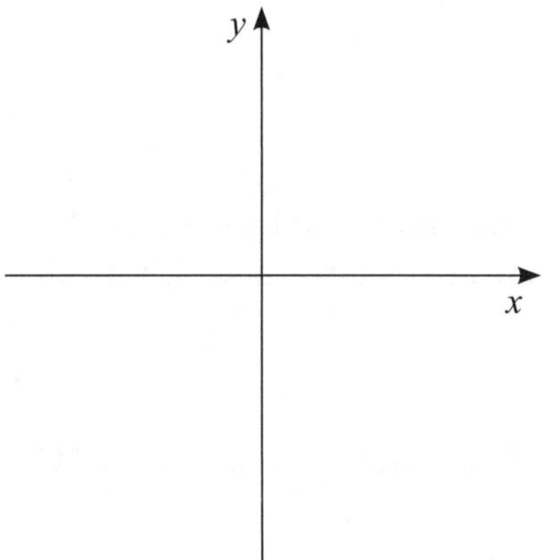

(5 marks)

8 Draw and label the region that satisfies the inequalities $y \geq x + 2$ and $4 - x^2 > y$.

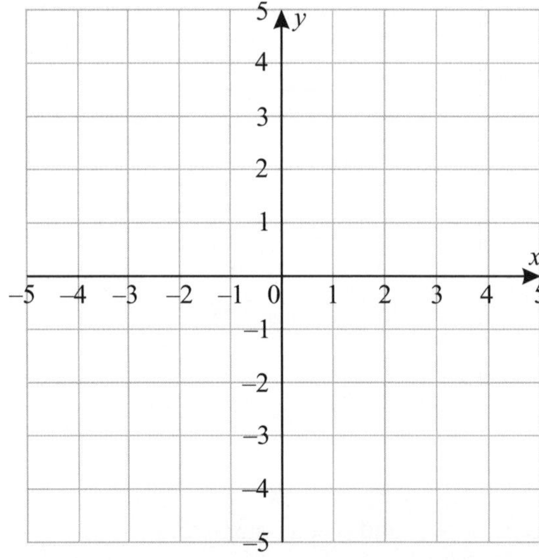

(3 marks)

9 Solve the inequality $x^2 - 8x + 15 > 0$. Give your answer in set notation.

....

(4 marks)

Algebra and Functions 2

10 $(x-1)(x^2+x+1) = 2x^2 - 17$

a) Rewrite the equation above in the form $f(x) = 0$, where $f(x)$ is of the form $f(x) = ax^3 + bx^2 + cx + d$.

....

(2 marks)

b) Show that $(x + 2)$ is a factor of $f(x)$.

(2 marks)

c) Hence write $f(x)$ as the product of a linear factor and a quadratic factor.

....

(2 marks)

d) By completing the square, or otherwise, show that $f(x) = 0$ has only one root.

(2 marks)

11 A function is defined by $f(x) = x^3 - 4x^2 - ax + 10$. $(x - 1)$ is a factor of $f(x)$.
Find the value of a and hence or otherwise solve the equation $x^3 - 4x^2 - ax + 10 = 0$.

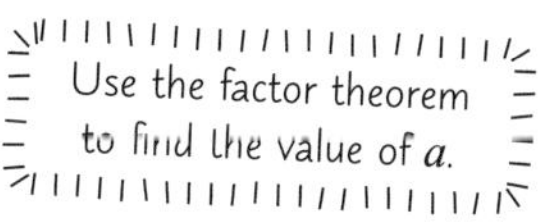

a = ... x = ...

(6 marks)

Quadratics have a habit of popping up in exam questions where you least expect them (like in exponentials, trig equations or mechanics, not to mention in simultaneous equations, inequalities and cubics) — so make sure you can handle them. It's worth practising your factorising skills, as it could save you a lot of time in the exam, and time is ~~money~~ marks.

Score

56

Algebra and Functions 3

This section should appeal to your artistic side — you get to draw some beautiful graphs. Although don't panic if you're not artistic — there's some equally beautiful algebra coming up later in the form of the binomial expansion.

1 $f(x) = |2x + 3|$ and $g(x) = |5x - 4|$.

a) On the same axes, draw the graphs of $y = f(x)$ and $y = g(x)$, showing clearly where each graph meets the coordinate axes.

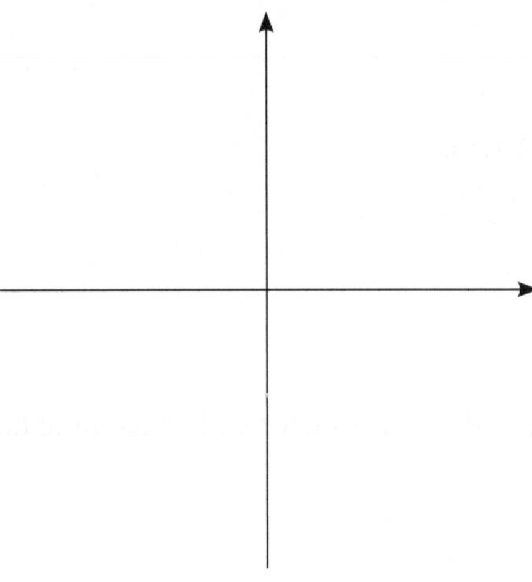

(2 marks)

b) Hence or otherwise solve the equation $f(x) = g(x)$.

....

(4 marks)

2 $f(x) = |4x + 5|$.

a) Find the possible values of $f(x)$ if $|x| = 2$.

....

(3 marks)

b) Find the values of x for which $f(x) \leq 2 - x$.

....

(3 marks)

c) Find the possible values of a constant A for which the equation $f(x) + 2 = A$ has two distinct roots.

....

(2 marks)

Algebra and Functions 3

3 Sketch the curve of $y = (x-2)^2(x+3)^2$ on the axes provided. Show clearly any points of intersection with the x- and y-axes.

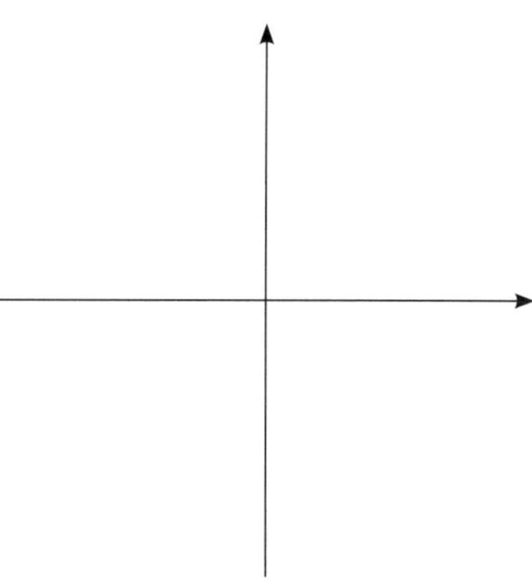

(3 marks)

4 The graph of $y = x^3 - 2x^2 + px$, for some constant p, crosses the x-axis at the points $(1-\sqrt{3}, 0)$, $(0, 0)$ and $(1+\sqrt{3}, 0)$.

a) Sketch the graph of $y = x^3 - 2x^2 + px$, showing clearly any points of intersection with the axes.

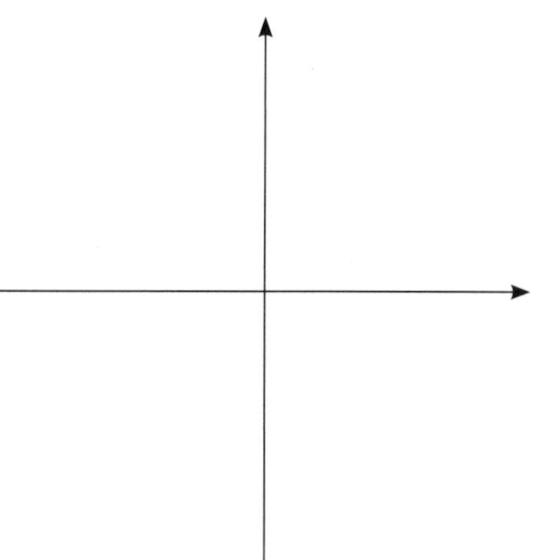

(2 marks)

b) Find the value of p.

It'll help to factorise the original function first.

....

(2 marks)

5 Describe what happens to the curve $y = x^3$ to transform it into the curve $y = 2(x-1)^3 + 4$.

....

....

....

(3 marks)

Algebra and Functions 3

6 A diver's position is modelled by the function $V = 2t^3 - 10t^2 + 8.5t + 7$, $0 \leq t \leq 3.5$.
V is the vertical height of the diver's head above the surface of the pool, in metres.
t is the time in seconds from him starting his dive.
At the deepest point of the dive, the diver is 3.70 m (3 s.f.) below the surface of the pool.

a) Show that V can be written as $V = (2t + 1)(t - 2)(t - 3.5)$.

(2 marks)

b) Hence sketch the graph of $y = V(t)$ for $0 \leq t \leq 3.5$.
Label any points of intersection between the graph and the axes.

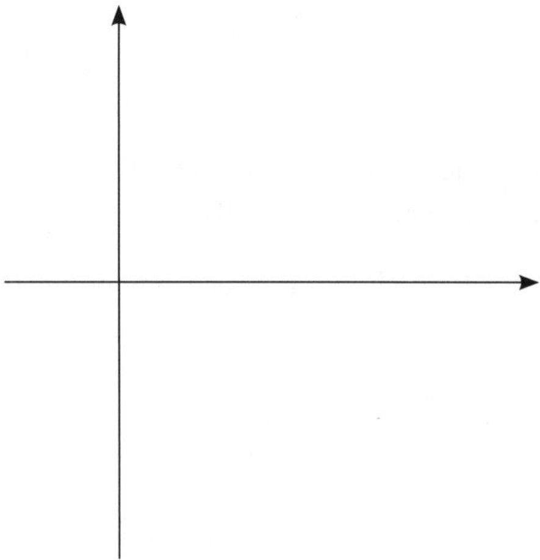

(3 marks)

c) How many seconds after starting the dive does the diver enter the pool?

.... s

(1 mark)

d) Given that the diver is 1.75 m tall, find the height of the diving board above the surface of the pool.

.... m

(2 marks)

e) The same diver then dives from a higher diving board. Darren suggests adapting the model for this diving board. The adapted model is $V = 2t^3 - 10t^2 + 8.5t + 10$.
Comment on the validity of this adapted model, giving reasons for your comments.

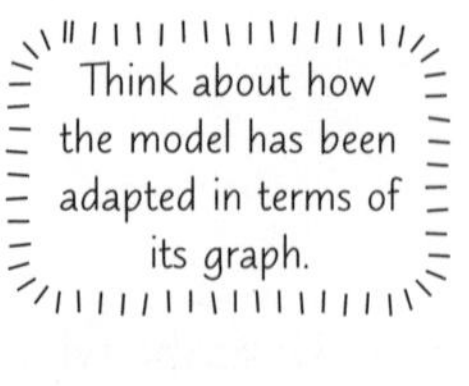

....

....

....

(2 marks)

Algebra and Functions 3

7 Figure 1 shows a sketch of the function $y = \mathrm{f}(x)$.
The function crosses the x-axis at $(-1, 0)$, $(1, 0)$ and $(2, 0)$, and crosses the y-axis at $(0, 2)$.

Figure 1

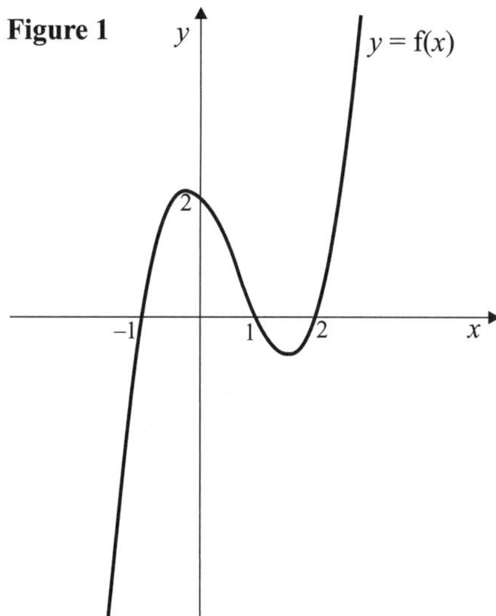

Sketch each transformation on the axes given below.
On each diagram, label any known points of intersection with the x- or y-axes.

a) $y = \mathrm{f}\left(\frac{1}{2}x\right)$

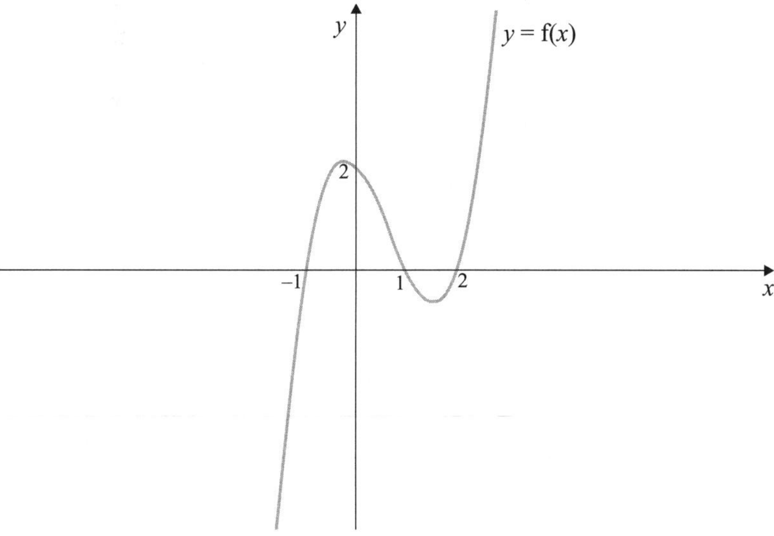

(3 marks)

b) $y = 2\mathrm{f}(x - 4)$

Do one transformation at a time to help keep track of what's going on.

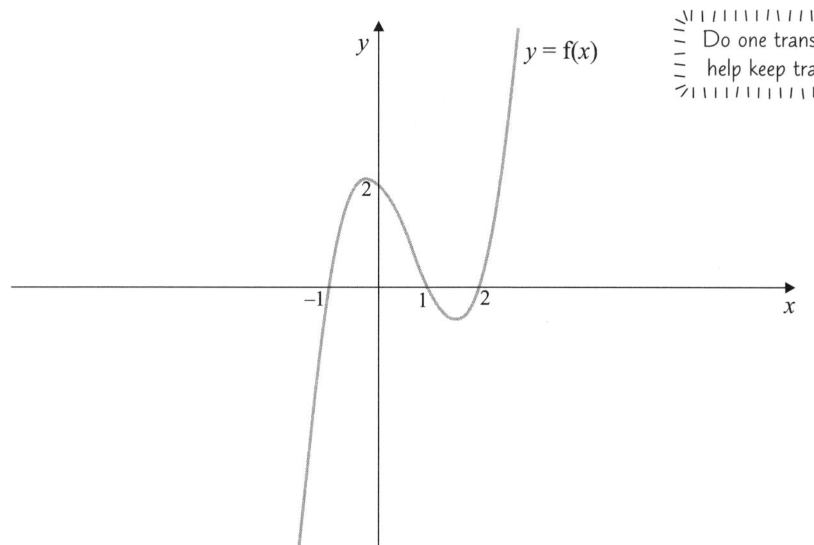

(3 marks)

Algebra and Functions 3

8 The graph on the right shows the function $y = f(x)$, $x \in \mathbb{R}$, with turning points $A(-1, -2)$ and $B(3, 2)$.

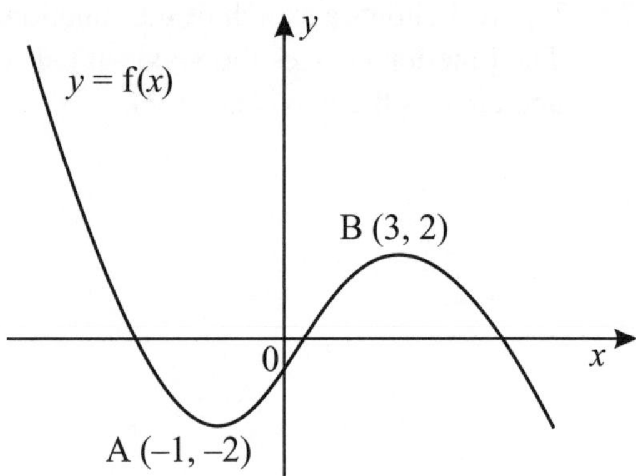

On separate axes, sketch the graphs of the following, clearly showing the coordinates of A and B where possible.

a) $y = f(-x) - 1$.

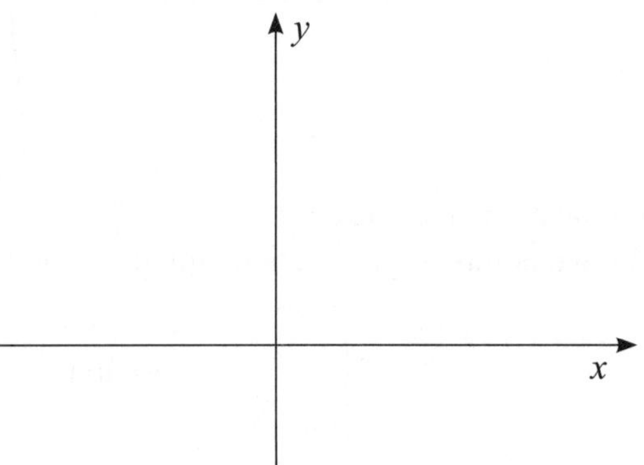

(3 marks)

b) $y = 3f(x + 2)$.

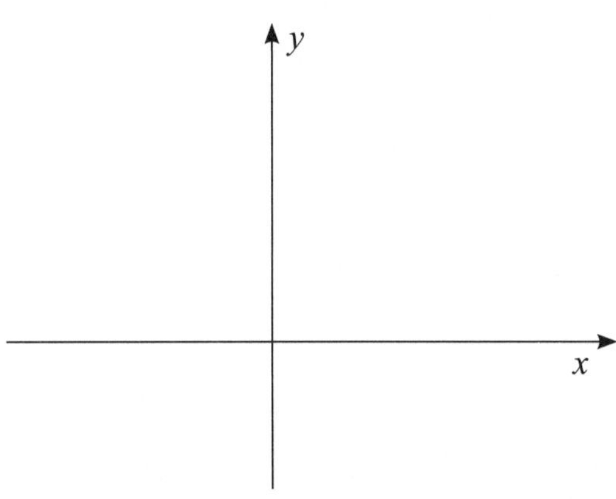

(3 marks)

9 The graph of the function $f(x)$ passes through points P(1, 2) and Q(3, 16). $f(x)$ is first translated 3 units up, then 2 units right and finally reflected in the y-axis to form a new graph, $g(x)$.

a) Write down the equation of $g(x)$ in terms of $f(x)$.

....

(3 marks)

b) Write down the new coordinates of the points P and Q.

....

(2 marks)

Algebra and Functions 3

10 The functions f and g are defined as follows: $f(x) = 2^x$, $x \in \mathbb{R}$ and $g(x) = \sqrt{3x+1}$, $x \geq -\frac{1}{3}$.

a) **(i)** Find gf(x).

....

(1 mark)

(ii) Hence solve gf(x) = 5.

....

(2 marks)

b) Find $g^{-1}(x)$ and state its domain and range.

....

(3 marks)

11 The functions f and g are defined as follows: $f(x) = \frac{1}{x^2}$, $x \in \mathbb{R}$, $x \neq 0$ and $g(x) = x^2 - k$, $x \in \mathbb{R}$, where k is a positive integer.

a) State the range of g. Give your answer in terms of k.

....

(1 mark)

b) Neither f nor g have an inverse. Explain why.

....

(1 mark)

c) **(i)** Given that gf(1) = –8, find the value of k and hence find fg(x), and write down the domain of the composite function fg.

....

(5 marks)

(ii) Hence solve $fg(x) = \frac{1}{256}$.

....

(4 marks)

Don't worry, graph sketches don't have to be perfect — as long as they're generally the correct shape, and any turning points and intersections are in the right places (and labelled if the question asks for it), you should get all the marks. It's always a good idea to sketch graphs in pencil in case you make a mistake (and make sure you have a rubber to hand just in case).

Score

68

Coordinate Geometry

Up, down, up, triangle, circle, circle, down, triangle... I'm afraid when it comes to coordinate geometry and circle equations, there are no cheat codes. So pens at the ready — you've got to do this the hard way.

1 The point A lies at the intersection of the lines l_1 and l_2, where the equation of l_1 is $x - y + 1 = 0$ and the equation of l_2 is $2x + y - 8 = 0$.

a) Find the coordinates of point A.

....

(3 marks)

b) The points B and C have coordinates (6, –4) and $\left(-\frac{4}{3}, -\frac{1}{3}\right)$ respectively, and D is the midpoint of AC. Find the equation of the line through B and D in the form $ax + by + c = 0$, where a, b and c are integers.

....

(5 marks)

c) Show that the triangle ABD is a right-angled triangle.

(3 marks)

2 The diagram shows a square ABCD, where point B has coordinates $(3, k)$. The line through points B and C has equation $-3x + 5y = 16$.

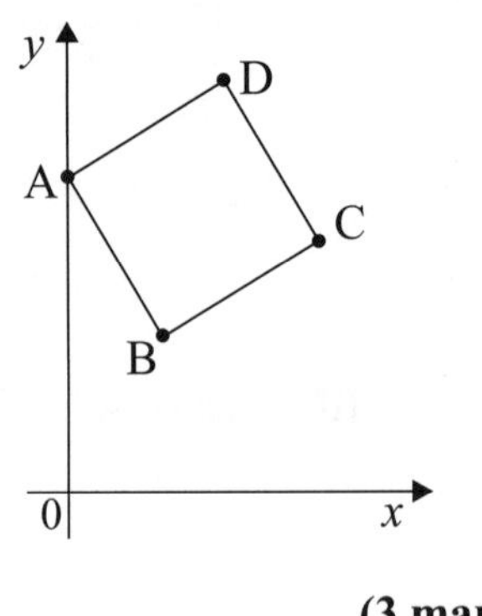

a) Show that the line with equation $5x + 3y - 6 = 0$ is parallel to the line through points A and B.

(3 marks)

b) Find the area of square ABCD.

....

(5 marks)

Coordinate Geometry

3 The points A(2, 1) and B(0, –5) lie on a circle, where the line AB is a diameter of the circle.

a) Find the centre and radius of the circle.

centre = ..., radius = ...

(3 marks)

b) Show that the point (4, –1) also lies on the circle.

(2 marks)

c) Show that the equation of the circle can be written in the form $x^2 + y^2 - 2x + 4y - 5 = 0$.

(2 marks)

d) Find the equation of the tangent to the circle at point A, giving your answer in the form $y = mx + c$.

....

(3 marks)

4 The diagram shows a circle with centre P.
The line AB is a chord with midpoint M.

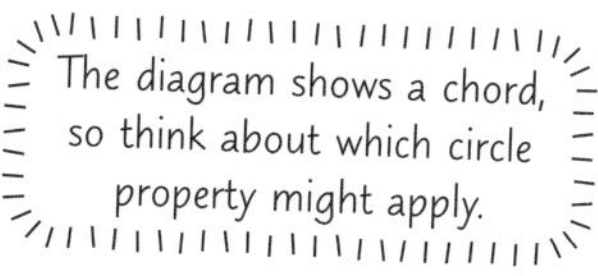

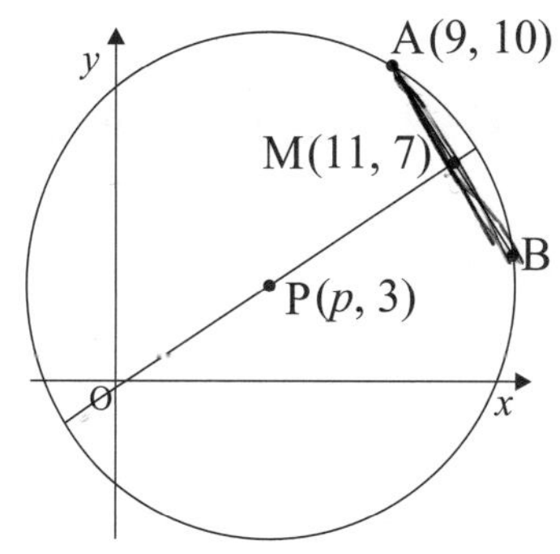

a) Show that $p = 5$.

(5 marks)

b) Find the equation of the circle.

....

(3 marks)

Coordinate Geometry

5 The circle with equation $x^2 - 6x + y^2 - 4y = 0$ crosses the y-axis at the origin and the point A.

a) Find the coordinates of point A.

....

(2 marks)

b) Write the equation of the circle in the form $(x - a)^2 + (y - b)^2 = c$.

....

(3 marks)

c) Write down the radius and the coordinates of the centre of the circle.

radius = ..., centre = ...

(2 marks)

d) The tangent to the circle at point A meets the x-axis at point B. Find the exact distance AB.

....

(6 marks)

6 PQR is a triangle whose vertices all lie on the circumference of circle C, where $P = (3, 1)$, $Q = (0, 2)$ and $R = (1, 5)$. Find the equation of C. Give your answer in the form $(x - a)^2 + (y - b)^2 = c$.

The perpendicular bisectors of the sides of the triangle will meet at the centre of the circle.

....

(7 marks)

Coordinate Geometry

7 The curve below has parametric equations $x = 3\sin\theta$, $y = 4\cos\theta$.

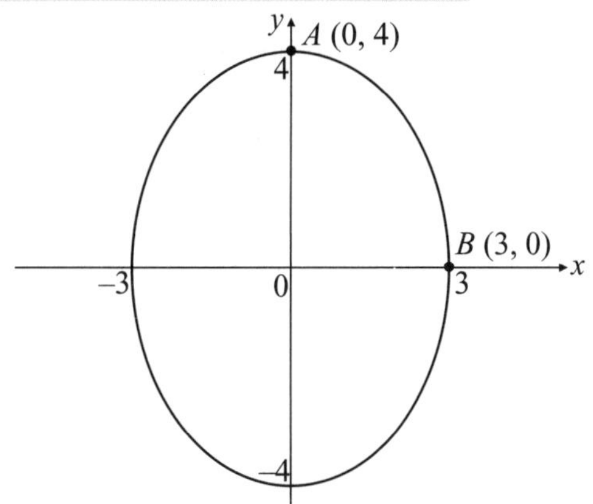

a) For $0 \leq \theta \leq \frac{\pi}{2}$, find the values of θ that correspond to the points A and B.

....

(2 marks)

b) Show that $y^2 = (4 + \frac{4x}{3})(4 - \frac{4x}{3})$.

You'll need to use a trig identity for this one...

(3 marks)

8 Part of the path of a boat sailing around an island is modelled on a shipping chart with the parametric equations $x = t^2 - 7t + 12$, $y = t - 1$, where t is time in hours.

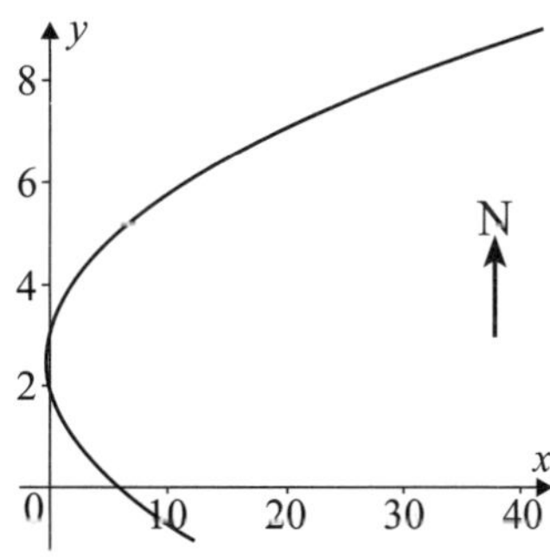

In this model, the western tip of the island has coordinates (12, 2).
Calculate the length of time that the boat is further west than the tip of the island.

.... hours

(3 marks)

Coordinate Geometry

9 The curve C is defined by the parametric equations:

$$x = 1 - \tan\theta,\ y = \frac{1}{2}\sin 2\theta, \quad -\frac{\pi}{2} < \theta < \frac{\pi}{2}.$$

a) P is the point on curve C where $\theta = \frac{\pi}{3}$. Find the exact coordinates of P.

....

(2 marks)

b) Point Q on curve C has coordinates $(2, -\frac{1}{2})$. Find the value of θ at Q.

....

(2 marks)

c) Using the identity $\sin 2\theta \equiv \frac{2\tan\theta}{1 + \tan^2\theta}$, show that the Cartesian equation of C is $y = \frac{1 - x}{x^2 - 2x + 2}$.

....

(3 marks)

10 A curve C has parametric equations $y = t^3 + t$, $x = 4t - 2$.

a) Find the coordinates of the point where the curve C crosses the y-axis.

....

(2 marks)

b) Find the coordinates of the points where the curve C intersects the line $y = \frac{1}{2}x + 1$.

....

(4 marks)

c) Find the Cartesian equation of the curve C in the form $y = ax^3 + bx^2 + cx + d$, where a, b, c and d are fractions.

Rearrange the equation for x to get an expression in terms of t.

$y =$ ....

(3 marks)

Working with parametric equations can be a real chore. When converting to Cartesian form, it's not always obvious how to rearrange the parametric equations, so you might have to try a couple of different ways — it's best to start with the equation that gives the simplest expression for t. Also, make sure you're up to snuff with your trig identities, as they'll come in handy.

Score

81

Sequences and Series 1

Time to test your knowledge of arithmetic and geometric sequences. Get ready for common differences, common ratios, recurrence relations, sums to infinity, sums to not-quite-infinity, sigma notation... I know you're gonna love it.

1 An arithmetic series has first term a and common difference d.
The value of the 12th term is 79, and the value of the 16th term is 103.

a) Find the values of a and d.

$a =$ .... $d =$ ....

(3 marks)

S_n is the sum of the first n terms of the series.

b) Find the value of S_{15}.

The formula for S_n is given on the formula sheet (see p.164-166).

....

(2 marks)

2 A sequence is defined by the recurrence relation $x_{n+1} = 3x_n - 4$, with first term $x_1 = k$.

a) Write a simplified expression for x_4 in terms of k.

....

(2 marks)

b) Determine whether the sequence is increasing or decreasing when $k = 1$.

(2 marks)

3 A sequence is defined by the recurrence relation: $h_{n+1} = 2h_n + 2$ when $n \geq 1$.

a) Given that $h_1 = 5$, find the values of h_2, h_3, and h_4.

....

(2 marks)

b) Calculate the value of $\sum_{r=3}^{6} h_r$.

....

(3 marks)

Sequences and Series 1

4 A geometric series $u_1 + u_2 + ... + u_n$ has 3rd term $\frac{5}{2}$ and 6th term $\frac{5}{16}$.

a) Find the formula for the nth term of the series.

....

(4 marks)

b) Find $\sum_{i=1}^{10} u_i$. Give your answer as a fraction in its simplest terms.

....

(2 marks)

c) Show that the sum to infinity of the series is 20.

(2 marks)

5 A geometric series has the first term 5 and is defined by: $u_{n+1} = 5 \times 1.7^n$.

a) Can the sum to infinity of this sequence be found? Explain your answer.

....

(1 mark)

b) Find the value of the 3rd and 8th terms.

$u_3 =$ .... $u_8 =$ ....

(2 marks)

6 In a geometric series with n^{th} term u_n, $a = 20$ and $r = \frac{3}{4}$.

Find values for the following, giving your answers to 3 significant figures where necessary:

a) S_∞

....

(2 marks)

b) u_{15}

....

(2 marks)

c) the smallest value of n for which $S_n > 79.76$.

Use the general formula for S_n, then start rearranging. You'll need to take logs to get n on its own.

....

(5 marks)

Sequences and Series 1

7 To raise money for charity, Alex, Chris and Heather were sponsored £1 for each kilometre they ran over a 10-day period. They receive sponsorship proportionally for partial kilometres completed.

Alex ran 3 km every day.

Chris ran 2 km on day 1 and on each subsequent day ran 20% further than the day before.

Heather ran 1 km on day 1, and on each subsequent day ran 50% further than the previous day.

a) How far did Heather run on day 5, to the nearest 10 metres?

.... km

(2 marks)

b) Show that day 10 is the first day that Chris runs further than 10 km.

(3 marks)

c) Find the total amount raised by the end of the 10 days, to the nearest penny.

(4 marks)

8 $a + ar + ar^2 + ar^3 + \ldots$ is a geometric series.
The second term of the series is -2 and the sum to infinity of the series is -9.

a) Show that $9r^2 - 9r + 2 = 0$.

(3 marks)

b) Find the possible values of r.

....

(2 marks)

c) Hence find the possible values of a.

....

(2 marks)

This is a top tip straight from the mouth of the examiners themselves, so you'd best listen up. There are a few different formulae to contend with here, so make sure you don't get flustered in the exam and get them muddled up. Don't get your u_n and your S_n mixed up or accidentally find the sum to infinity when you don't mean to — you're just throwing away marks that way.

Score

50

Sequences and Series 2

OK, now it's time for something a little different. The next few pages cover variations of binomial expansions — including cases where the power is a positive integer, a negative integer or even a fraction.

1 Use the binomial expansion formula to answer the following questions.

a) Write down the first four terms in the expansion of $(1 + ax)^{10}$, $a > 0$.

....

(2 marks)

b) Find the coefficient of x^2 in the expansion of $(2 + 3x)^5$.

....

(3 marks)

c) Given that the coefficients of x^2 in both expansions are equal, find the value of a.

$a =$ ....

(1 mark)

2 The binomial expansion of $(j + kx)^6$ is $j^6 + ax + bx^2 + cx^3 + ...$

a) Given that $c = 20\ 000$, show that $jk = 10$ (where both j and k are positive integers).

(3 marks)

b) Given that $a = 37\ 500$, find the values of j and k.

$j =$ ....

$k =$ ....

(4 marks)

c) Find b.

$b =$ ....

(2 marks)

Sequences and Series 2

3 $(1-x)^{-\frac{1}{2}} \approx 1 + Ax + Bx^2 + Cx^3$, where A, B and C are rational constants.

a) Find the values of the constants A, B and C.

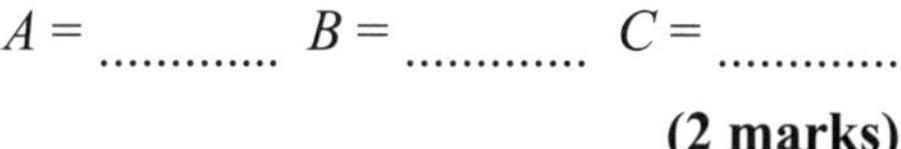

(2 marks)

b) **(i)** Hence show that $(25-4x)^{-\frac{1}{2}} \approx \frac{1}{5} + \frac{2}{125}x + \frac{6}{3125}x^2 + \frac{4}{15\,625}x^3$ for small values of x.

(4 marks)

(ii) State the range of values of x for which the expansion from part (i) is valid.

....

(1 mark)

c) **(i)** Use your expansion from b) with a suitable value of x to show that $\frac{1}{\sqrt{20}} \approx \frac{447}{2000}$.

(3 marks)

(ii) What is the percentage error in this estimate? Give your answer to 3 significant figures.

.... %

(2 marks)

Sequences and Series 2

4 $f(x) = (27 + 4x)^{\frac{1}{3}}$, for $|x| < \frac{27}{4}$

a) Find the binomial expansion of f(x) up to and including the term in x^2.

....

(4 marks)

b) Hence find an approximation to $\sqrt[3]{26.2}$. Give your answer to 6 decimal places.

....

(2 marks)

5 $f(x) = \sqrt{\frac{1+3x}{1-5x}}$

a) (i) Use the binomial expansion in increasing powers of x to show that $f(x) \approx 1 + 4x + 12x^2$.

Split f(x) up into two binomial expansions and work them out separately — then combine your expansions at the end.

(5 marks)

(ii) For what values of x is your expansion valid?

....

(2 marks)

b) Using the above expansion with $x = \frac{1}{15}$, show that $\sqrt{1.8} \approx \frac{33}{25}$.

(2 marks)

Sequences and Series 2

6 $f(x) = \dfrac{2 - 18x}{(5 + 4x)(1 - 2x)^2}$

a) Given that f(x) can be expressed in the form $f(x) = \dfrac{A}{(5 + 4x)} + \dfrac{B}{(1 - 2x)} + \dfrac{C}{(1 - 2x)^2}$, find the values of A, B and C.

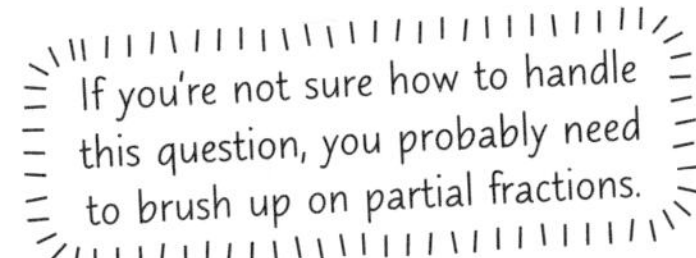

$A =$ .... $B =$ .... $C =$ ....

(4 marks)

b) Hence find the binomial expansion of f(x), up to and including the term in x^2.

....

(7 marks)

c) Show that the expansion above is valid for values of x in the range $-\frac{1}{2} < x < \frac{1}{2}$.

(2 marks)

The binomial expansion formulas are given in the exam (one for positive integer n and one for any real value of n) — make sure you know when to use each one and how they work. As well as being happy with doing the expansions, make sure you're confident with the other things examiners often ask — such as approximations and ranges of validity.

Score

55

Trigonometry

You'd think that there's only so much you can do with a three-sided shape, right? Well, think again...
It's time to delve deeper into the twisted triangular world of trig, and discover new ways of measuring angles.

1 The diagram shows a sector of a circle of radius r cm and angle 120°. The length of the arc of the sector is 40 cm.

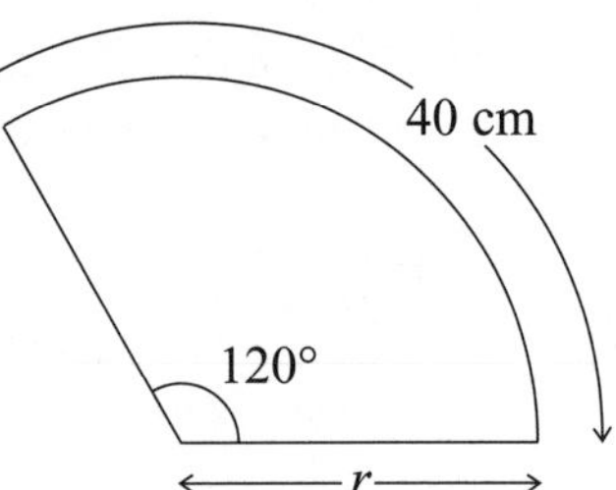

Find the area of the sector to the nearest square centimetre.

.... cm^2

(5 marks)

2 A new symmetrical mini-stage is to be built according to the design below. The design consists of a rectangle of length q metres and width $2r$ metres, two sectors of radius r and angle θ radians (shaded), and an isosceles triangle.

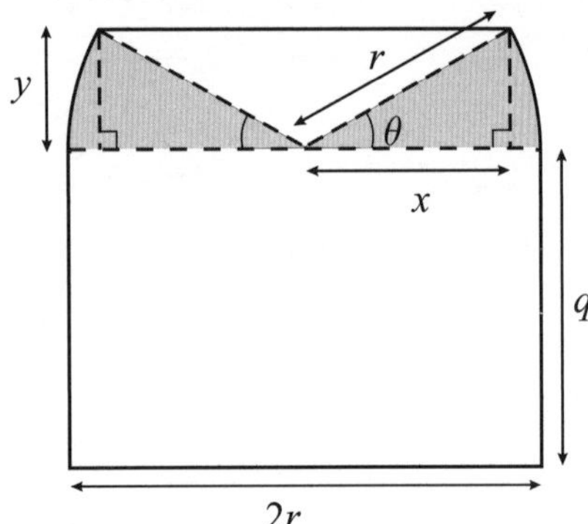

a) **(i)** Show that distance x is given by $x = r\cos\theta$.

(1 mark)

(ii) Find a similar expression for distance y.

....

(1 mark)

b) Find, in terms of r, q and θ, expressions for the perimeter P, and the area A, of the stage.

....

(4 marks)

c) Given that the perimeter of the stage is to be 40 m and $\theta = \frac{\pi}{3}$,

show that A is given by $A = 40r - kr^2$, where $k = 3 - \frac{\sqrt{3}}{4} + \frac{\pi}{3}$.

(4 marks)

Trigonometry

3 Sketch the graphs of $y = \sin x$ and $y = \sin \frac{x}{2}$ in the range $0 \leq x \leq 4\pi$ on the same set of axes, showing the points at which the graphs cross the x-axis.

(3 marks)

4 A sheep pen is modelled as a triangle with side lengths of 50 m, 70 m and 90 m.

a) Find the area of the sheep pen to the nearest square metre.

.... m^2

(5 marks)

b) Comment on the accuracy of the model.

....

....

(1 mark)

5 Solutions to this question based entirely on graphical or numerical methods are not acceptable.

Find all the values of x, in the interval $0° \leq x \leq 180°$, for which $7 - 3\cos x = 9\sin^2 x$.

....

(5 marks)

Trigonometry

6 Adam and Bethan have each attempted to solve the equation $\sin 2t = \sqrt{2}\cos 2t$ for the range $-90° < t < 90°$. Their working is shown below.

Adam

$\sin 2t = \sqrt{2}\cos 2t$

$\tan 2t = \sqrt{2}$

$\tan t = \frac{\sqrt{2}}{2}$

$t = 35.26...°$

Bethan

$\sin 2t = \sqrt{2}\cos 2t$

$\sin^2 2t = 2\cos^2 2t$

$1 - \cos^2 2t = 2\cos^2 2t$

$\cos^2 2t = \frac{1}{3}$

$\cos 2t = \pm\frac{1}{\sqrt{3}}$

$t = \pm 27.36...°$

a) Show that Adam's solution is incorrect.

(1 mark)

b) Identify an error made by Adam.

....

....

(1 mark)

c) Bethan's teacher explains that one of her solutions is incorrect.
Identify and explain the error Bethan has made.

....

....

(2 marks)

7 In this question, you must show detailed reasoning.

a) Show that the equation $\tan^2\theta + \frac{\tan\theta}{\cos\theta} = 1$ can be written in the form $2\sin^2\theta + \sin\theta - 1 = 0$.

(3 marks)

b) Hence find all solutions to the equation $\tan^2\theta + \frac{\tan\theta}{\cos\theta} = 1$ in the interval $0 \le \theta \le 2\pi$.

....

(4 marks)

Trigonometry

8 The diagram on the right shows the graph of $y = \arccos x$, where y is in radians. A and B are the end points of the graph.

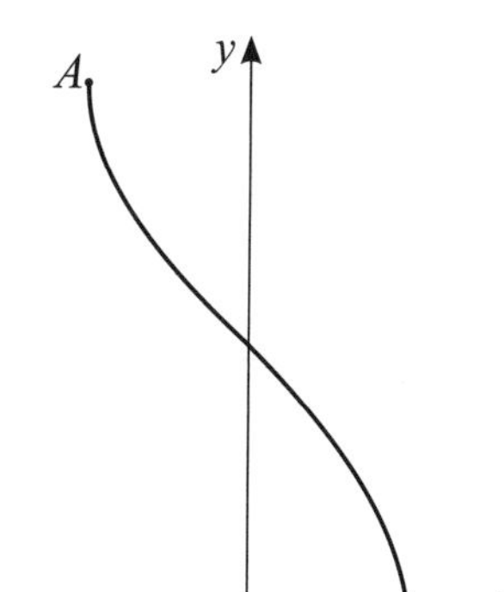

a) Write down the coordinates of A and B.

$A =$ .... , $B =$ ....

(2 marks)

b) Express x in terms of y.

....

(1 mark)

c) Solve, to 3 significant figures, the equation $\arccos x = 2$ for the interval shown on the graph.

....

(2 marks)

9 In this question, give your answers to 3 significant figures where appropriate.

a) Find the values of θ in the range $0 \leq \theta \leq 2\pi$ for which $\operatorname{cosec}\theta = \frac{5}{3}$.

....

(2 marks)

b) **(i)** Use an appropriate identity to show that $3\operatorname{cosec}\theta = \cot^2\theta - 17$ can be written as $18 + 3\operatorname{cosec}\theta - \operatorname{cosec}^2\theta = 0$.

(2 marks)

(ii) Hence solve the equation $3\operatorname{cosec}\theta = \cot^2\theta - 17$ for $0 \leq \theta \leq 2\pi$.

....

(5 marks)

10 Given that θ is small and measured in radians, show that $4\sin\theta\tan\theta + 2\cos\theta$ can be approximated by the expression $p + q\theta^2$, where p and q are integers to be found.

(3 marks)

Trigonometry

11 Given that $\cos x = \frac{8}{9}$ for the acute angle x, find the exact values of:

a) $\sec x$

(1 mark)

b) $\operatorname{cosec} x$

(2 marks)

c) $\tan^2 x$

(2 marks)

d) $\cos 2x$

$\cos 2x = \cos^2 x - \sin^2 x$

$2\cos^2 x - 1$

$\therefore 2\left(\frac{8}{9}\right)^2 - 1 = \frac{47}{81}$

(3 marks)

12 The graph below shows the curve of $y = \frac{1 + \cos x}{2}$:

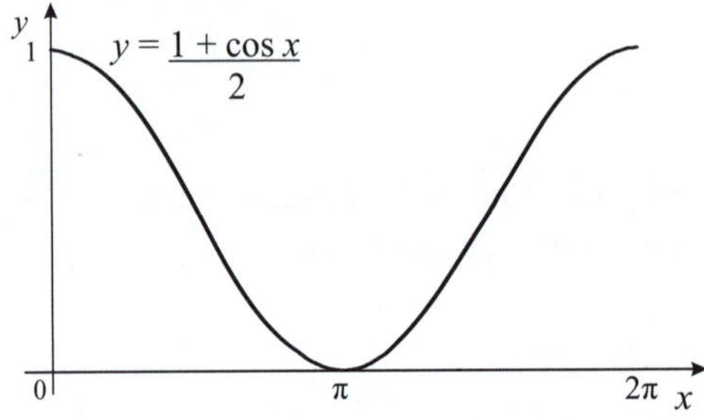

a) Show that $\frac{1 + \cos x}{2} = \cos^2 \frac{x}{2}$.

Use the double angle formula for cos.

(2 marks)

b) Hence find the exact values of x for which $\cos^2 \frac{x}{2} = 0.75$ in the interval $0 \leq x \leq 2\pi$.

(2 marks)

Trigonometry

13 By writing $\sin 2\theta$ in terms of $\sin\theta$ and $\cos\theta$, solve the equation $3\sin 2\theta \tan\theta = 5$, for $0 \leq \theta \leq 2\pi$. Give your answers to 3 significant figures.

....

(6 marks)

14 Figure 1 shows an isosceles triangle ABC with $AB = AC = 2\sqrt{2}$ cm and $\angle BAC = 2\theta$. The midpoints of AB and AC are D and E respectively. A rectangle $DEFG$ is drawn inscribed in the triangle, with F and G on BC. The perimeter of rectangle $DEFG$ is P cm.

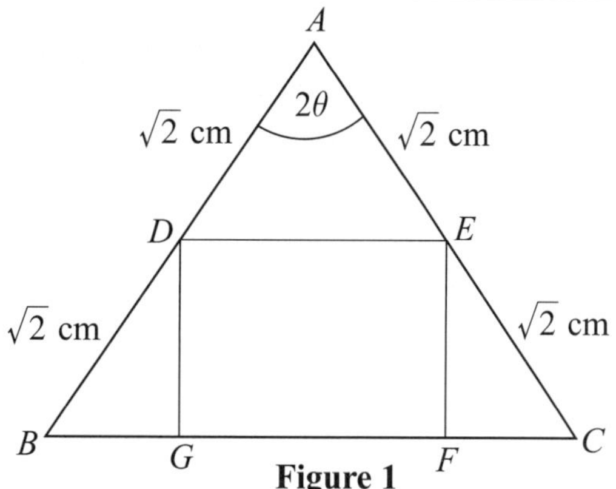

Figure 1

a) Using the cosine rule, $DE^2 = 4 - 4\cos 2\theta$.
Use a trigonometric identity to show that $DE = 2\sqrt{2}\sin\theta$.

(2 marks)

b) Show that $P = 4\sqrt{2}\sin\theta + 2\sqrt{2}\cos\theta$

(2 marks)

c) Express P in the form $R\sin(\theta + \alpha)$ where $R > 0$ and $0 < \alpha < \frac{\pi}{2}$.

....

(3 marks)

Trigonometry

15 A garden sprinkler system is set up to water a flower bed. The distance, d feet, the water sprays at time θ minutes can be modelled by the function $d = \sqrt{2}\cos\theta - 3\sin\theta$, where distance to the left of the sprinkler is modelled as negative and distance to the right of the sprinkler is modelled as positive.

a) Write $\sqrt{2}\cos\theta - 3\sin\theta$ in the form $R\cos(\theta + \alpha)$, where $R > 0$ and $0 \leq \alpha \leq \frac{\pi}{2}$.

....

(3 marks)

b) Find the times at which the water sprays 3 feet to the right of the sprinkler within the first 6 minutes after being switched on. Give your answers in minutes and seconds, to the nearest second.

....

(4 marks)

c) The distance (d) in feet that water is sprayed by an industrial sprinkler for a farmer's field is modelled by the function $d = (\sqrt{2}\cos\theta - 3\sin\theta)^4$. θ is the time in minutes.
Find the maximum distances to the left and right of this sprinkler that the water reaches.

....

(2 marks)

d) Give one possible explanation for the minimum distance found in part c) in the context of this model.

....

....

(1 mark)

16 Show that $2\tan A \operatorname{cosec} 2A \equiv 1 + 1\tan^2 A$.

(3 marks)

Be very careful with degrees and radians. If the question gives the range in radians, you must give your answer in radians, otherwise you'll lose marks. Make sure your calculator is set to radians when appropriate too. There are some bits of trig that you have to use radians for — like the small angle approximations and arc length and sector area. Better get used to them.

Score

95

Exponentials and Logarithms

Exponentials and logs might seem a bit tricky at first, but once you get used to them and they get used to you, you'll wonder what you ever worried about. Plus, they're really handy for modelling real-life situations.

1 Given that $p > 0$, what is the value of $\log_p(p^4) + \log_p(\sqrt{p}) - \log_p\left(\frac{1}{\sqrt{p}}\right)$?

....

(3 marks)

2 Solve the equation $5^{(z^2 - 9)} = 2^{(z - 3)}$, giving your answers to 3 significant figures where appropriate.

....

(5 marks)

3 Solve the equation $3^{2x} - 9(3^x) + 14 = 0$, giving each solution to an appropriate degree of accuracy.

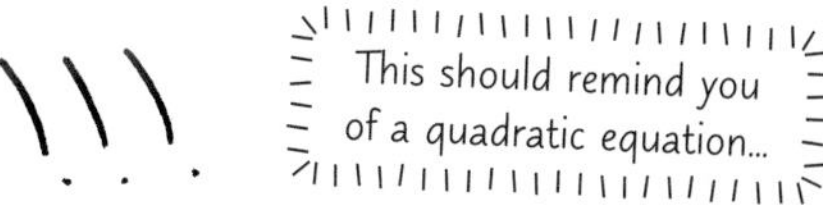

....

(5 marks)

4 The curve with equation $y = \ln(4x - 3)$ is shown on the graph to the right.

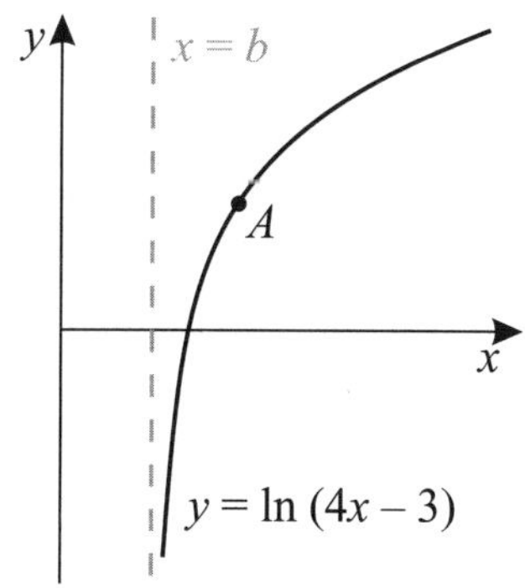

a) The point A with coordinates $(a, 1)$ lies on the curve. Find a to 2 decimal places.

$a =$ ....

(2 marks)

b) The curve only exists for $x > b$. State the value of b.

$b =$ ....

(2 marks)

Exponentials and Logarithms

5 The curve below has equation $y = Ae^{bx}$.

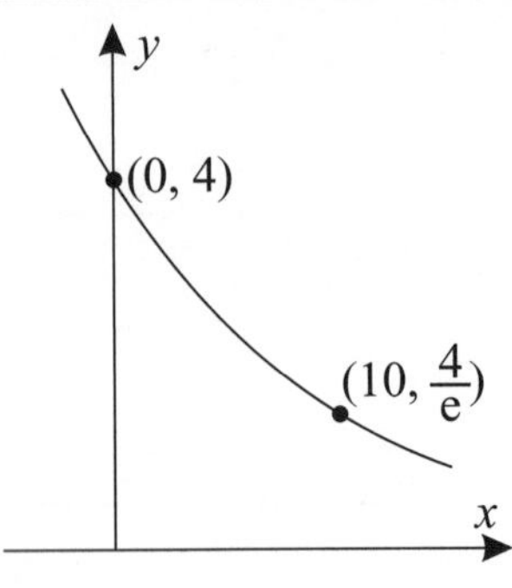

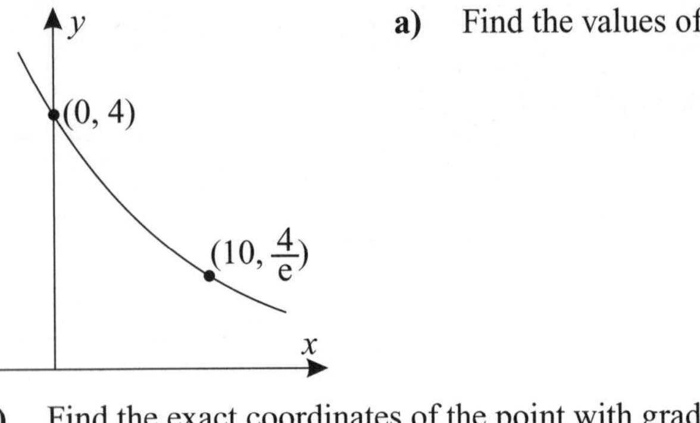

a) Find the values of A and b.

$A =$ .... $b =$ ....

(3 marks)

b) Find the exact coordinates of the point with gradient -1.

....

(5 marks)

6 The sketch on the right shows the function $y = e^{ax} + b$, where a and b are constants.

Find the values of a and b, and the equation of the asymptote shown on the sketch.

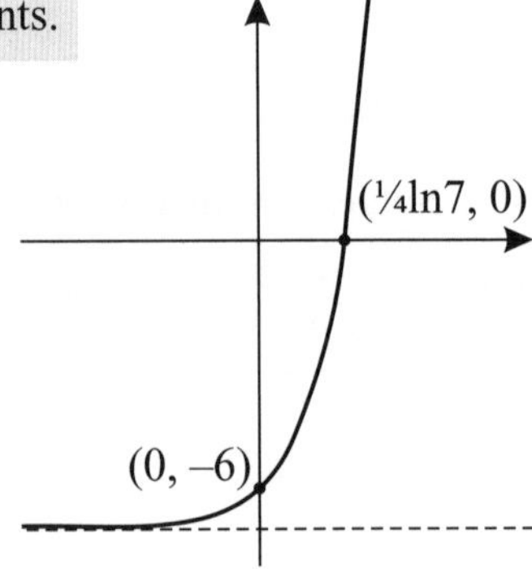

....

(4 marks)

7 Rudy buys a motorbike for £8000. The value of the motorbike, £m, after t months, is modelled by the function $m = m_0 e^{kt}$, where m_0 is the initial value of the motorbike. k is given by the formula $k = \frac{1}{12}\ln\left(1 - \frac{r}{100}\right)$, where r% is the rate of depreciation per year. For the first year, the motorbike depreciates by 8% per year, then the rate drops to 4% per year for each subsequent year. After how many months will the motorbike be worth less than half its original value?

You'll have to use a new value of m_0 after the first year (when the rate of depreciation changes).

....

(5 marks)

Exponentials and Logarithms

8 The UK population, P, of an endangered species of bird is modelled over time, t years ($t \geq 0$), by the function: $P = 5700e^{-0.15t}$. The UK population, Q, of a bird of prey that hunts the endangered species, as well as other animals, is modelled by the function: $Q = 2100 - 1500e^{-0.15t}$. The time $t = 0$ represents the beginning of the year 2010, when the bird of prey was first introduced into the country.

a) Find the year in which the population of the bird of prey is first predicted to exceed the population of the endangered species according to these models.

....

(4 marks)

b) The graph showing the predicted population of the bird of prey is shown on the right. Add a curve to the graph to show the predicted UK population of the endangered species of bird over the same time period.

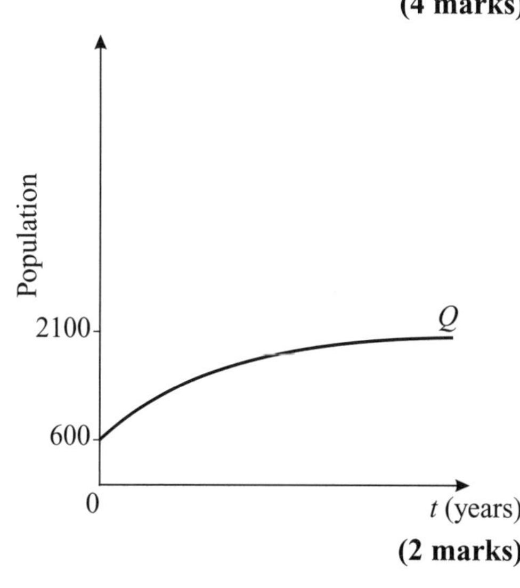

(2 marks)

c) Comment on the validity of each population model.

....

....

....

....

(2 marks)

d) Predict the year that the population of the endangered species will drop to below 1000.

....

(3 marks)

e) When this population drops below 1000, conservationists start enacting a plan to save the species. Suggest one refinement that could be made to the model to take this into account.

....

....

(1 mark)

Exponentials and Logarithms

9 The number of supporters of a local football team has tended to increase in recent years. The attendance can be modelled by an equation of the form $y = ab^t$, where y is the average home game attendance in hundreds, t is the number of years after the 2010/11 season, and a and b are constants to be determined.

a) Show that $y = ab^t$ can be written in the form $\log_{10}y = t\log_{10}b + \log_{10}a$.

(2 marks)

The graph of $\log_{10}y$ against t has been plotted below. A line of best fit has been drawn on the graph.

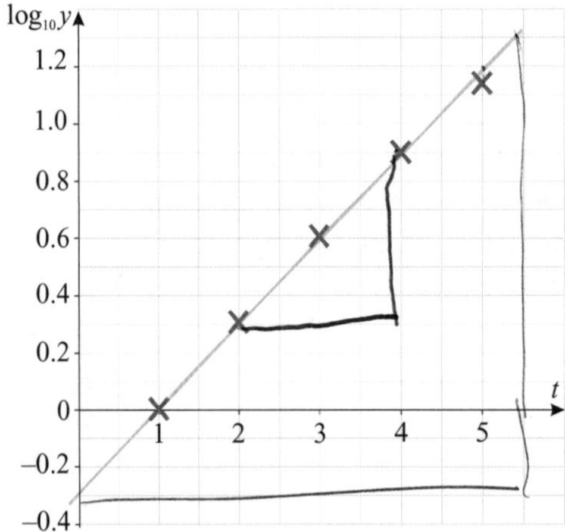

b) A new stadium will be built for the team when the average home game attendance exceeds 5000. Use the information provided to predict the season in which the attendance will reach this value.

....

(5 marks)

c) Interpret the value of a in the context of the model.

....

....

(1 mark)

d) Can this model be used to accurately predict the number of fans in the 2024/25 season? Explain your answer.

....

....

(2 marks)

If you're asked to comment on the validity of models, or asked for limitations or refinements, think about anything that stops the model being realistic. For example, it might predict that a quantity will continue to rise until it becomes infinitely big, which might not be likely in a real-life situation — so you could refine the model by introducing an upper limit.

Score

56

Differentiation 1

Differentiation can tell you all sorts of useful things about the gradient of a curve. The following questions use simple functions in powers of x — you'll find trickier functions in 'Differentiation 2'.

1 The curve C is given by the equation $y = 2x^3 - 10x^2 - 4\sqrt{x} + 12$.

Find the gradient of the tangent to the curve at the point where $x = 4$.

....

(4 marks)

2 For $f(x) = x^3 - 7x^2 + 8x + 9$, sketch the graph of $y = f'(x)$, showing clearly the points of intersection with the axes.

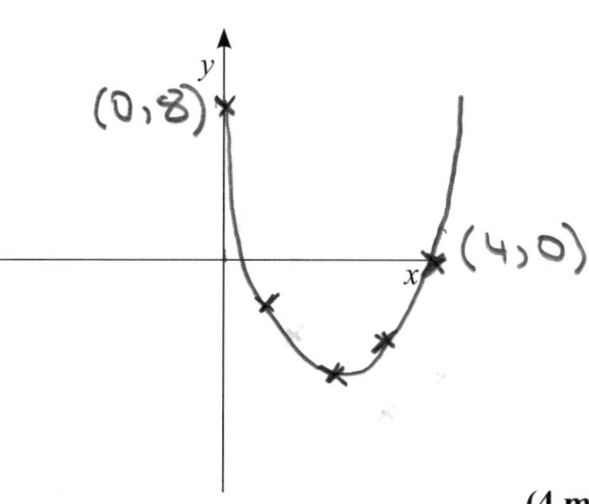

(4 marks)

3 A curve has equation $y = kx^2 - 8x - 5$, for a constant k. The point R lies on the curve and has an x-coordinate of 2. The normal to the curve at point R is parallel to the line with equation $4y + x = 24$.

a) Find the value of k.

$k =$ ....

(5 marks)

b) The tangent to the curve at R meets the curve $y = 4x - \frac{1}{x^3} - 9$ at the point S. Find the coordinates of S.

....

(5 marks)

Differentiation 1

4 For $f(x) = 8x^2 - 1$, prove from first principles that $f'(x) = 16x$.

(4 marks)

5 The function $f(x) = 2x^4 + 27x$ has one stationary point.

a) Find the coordinates of the stationary point.

....

(4 marks)

b) Find the range of values of x for which the function is increasing and the range of values of x for which it is decreasing.

Increasing for: ...., decreasing for: ....

(2 marks)

c) Hence sketch the curve $y = f(x)$, showing where it crosses the axes and the position of its stationary point.

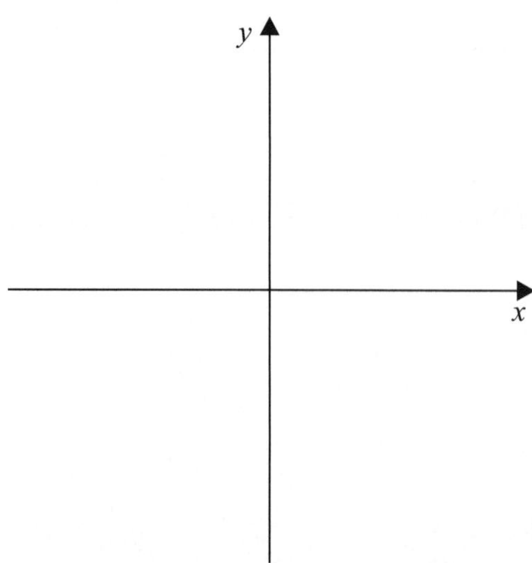

(3 marks)

Differentiation 1

6 The diagram shows part of the graph of $y = x^4 + 3x^3 - 6x^2$.
Find the range of values of x for which the graph is concave downwards.

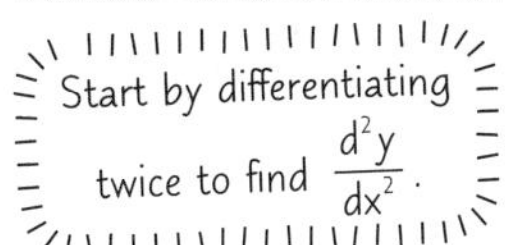

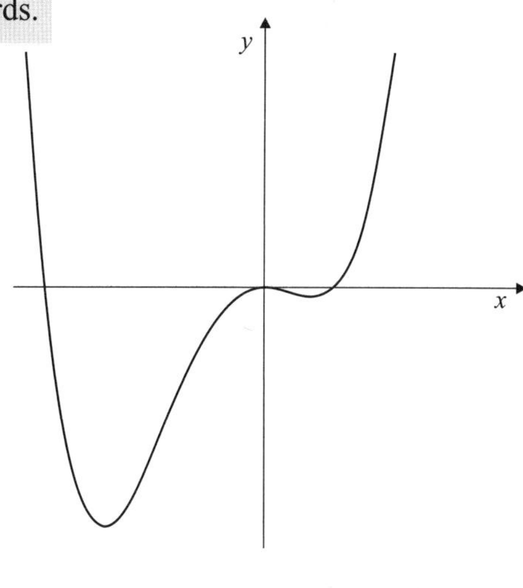

....

(4 marks)

7 The function $f(x) = 3x^3 + 9x^2 + 25x$ has one point of inflection.

a) Show that the point of inflection is at $x = -1$.

(5 marks)

b) Explain whether this is a stationary or non-stationary point of inflection.

....

....

(2 marks)

c) Joe claims that the function $f(x) = 3x^3 + 9x^2 + 25x$ is an increasing function for all values of x.
Show that Joe is correct.

(2 marks)

Differentiation 1

8 An ice cream parlour needs an open-top stainless steel container with a capacity of 40 litres, modelled as a cuboid with sides of length x cm, x cm and y cm, as shown in Figure 1.

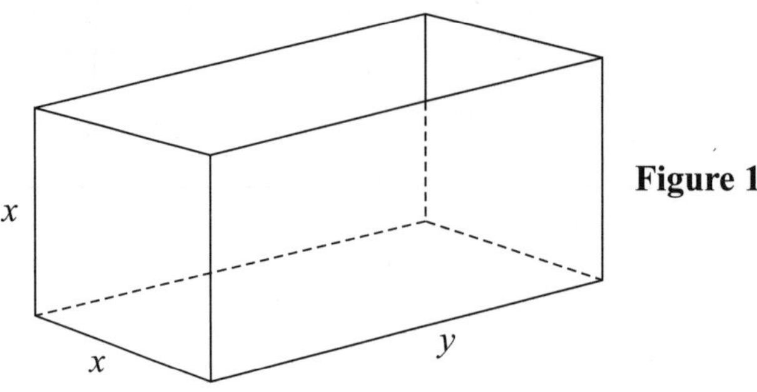

Figure 1

a) Show that the external surface area, A cm^2, of the container is given by $A = 2x^2 + \dfrac{120\,000}{x}$.

Find y in terms of x using the info given in the question.

(4 marks)

b) Find the value of x to 3 s.f. at which A is stationary, and show that this is a minimum value of A.

(6 marks)

c) Calculate the minimum area of stainless steel needed to make the container. Give your answer to 3 s.f.

.... cm^2

(2 marks)

d) Comment on the validity of this model.

....

....

(1 mark)

Remember, tangents have the same gradient as the curve and normals are perpendicular to the curve. To find stationary points, differentiate the expression and set equal to zero. And finally... $f''(x) > 0$ means the gradient is increasing (the curve is concave upwards), $f''(x) < 0$ means the gradient is decreasing (the curve is concave downwards) and $f''(x) = 0$ at a point of inflection.

Score

57

Differentiation 2

As promised, more differentiation. Make sure you use the correct rule for each type of function and you'll be fine.

1 Find $\frac{dy}{dx}$ at the given point for each of the following:

a) $y = \frac{1}{\sqrt{2x - x^2}}$, (1, 1)

....

(4 marks)

b) $x = (4y + 10)^3$, (8, –2)

....

(4 marks)

2 The diagram shows a container in the shape of a regular tetrahedron, which is being filled with water. After t minutes, the water in the container can be modelled as a regular tetrahedron with edge length a cm and vertical height x cm, and the volume of water in the container is V cm^3.

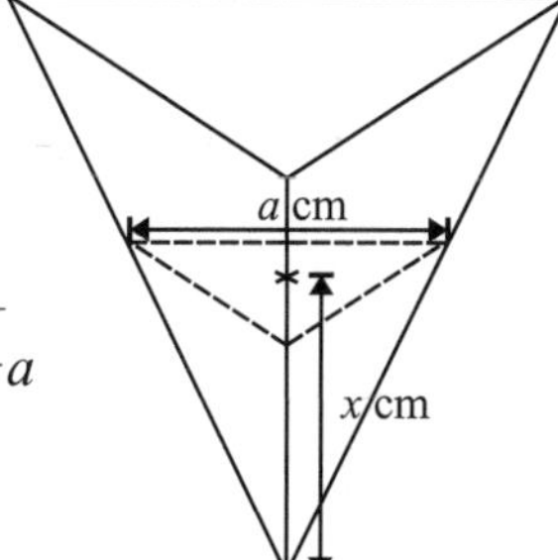

a) Given that a regular tetrahedron with edge length a has vertical height $h = \sqrt{\frac{2}{3}}a$ and volume $\frac{\sqrt{2}}{12}a^3$, show that the volume of water in the container after t minutes is given by $V = \frac{\sqrt{3}}{8}x^3$.

(2 marks)

Water is being poured into the container at a constant rate of 240 cm^3 min^{-1}.

b) Find the exact value of $\frac{dx}{dt}$ when the vertical height of the water in the container is 8 cm. Give units with your answer.

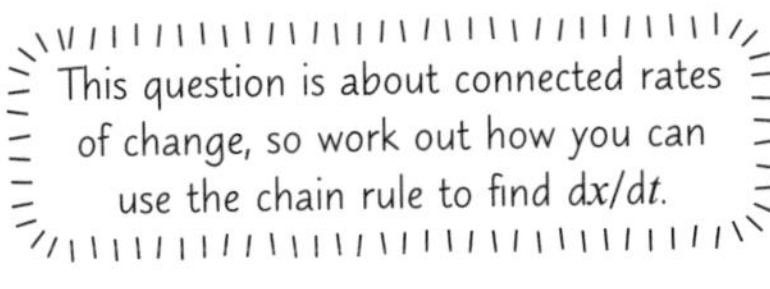

....

(5 marks)

The value of $\frac{dx}{dt}$ is measured when $x = 12$ and found to be $\frac{32}{9\sqrt{3}}$. This is less than the value expected if $\frac{dV}{dt} = 240$ cm^3 min^{-1}. It is discovered that water has been leaking out of the container at a constant rate.

c) Find $\frac{dV}{dt}$ if $\frac{dx}{dt} = \frac{32}{9\sqrt{3}}$ when $x = 12$. Give units with your answer.

....

(3 marks)

Differentiation 2

3 A curve has the equation $y = e^{2x} - 5e^{x} + 3x$.

a) Find $\frac{dy}{dx}$.

....

(2 marks)

b) Show that the stationary points on the curve occur when $x = 0$ and $x = \ln \frac{3}{2}$.

(4 marks)

c) Determine the nature of each of the stationary points.

(3 marks)

4 The curve C is given by the equation $y = 2^{3x}$.

a) Find the exact value of $\frac{dy}{dx}$ at the point on the curve where $x = 1$.

....

(3 marks)

b) Explain why $y = 2^{3x}$ is an increasing function for all values of x.

....

....

....

(2 marks)

Differentiation 2

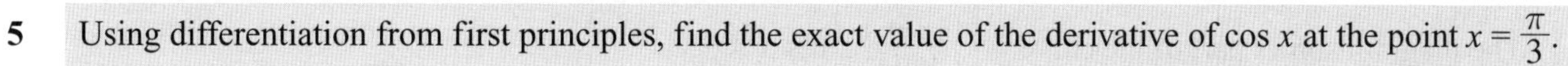

5 Using differentiation from first principles, find the exact value of the derivative of $\cos x$ at the point $x = \frac{\pi}{3}$.

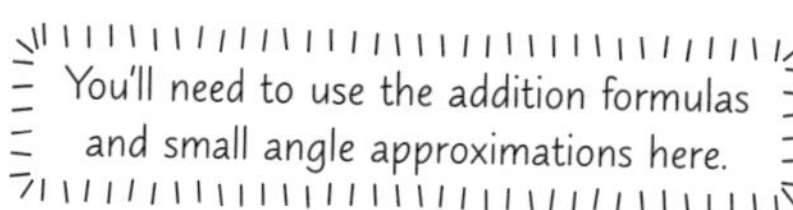

....

(6 marks)

6 For $y = \ln x\,(5x - 2)^3$, show that $\frac{dy}{dx} = (5x - 2)^2\left(a\ln x + b + \frac{c}{x}\right)$,
where a, b and c are constants to be found.

(4 marks)

7 A curve has the equation $y = \frac{5 - 2x}{3x^2 + 3x}$.

Show that, at the stationary points on the curve, $2x^2 - 10x - 5 = 0$.

(6 marks)

8 A curve has the equation $y = 4x^2 \ln x, \quad x > 0$.

Find the ranges of values of x for which the curve is concave downwards and concave upwards.

You need to use the product rule twice.

Concave downwards for: ...., Concave upwards for: ....

(8 marks)

Differentiation 2

9 The diagram shows part of the curve $y = \dfrac{4x-1}{\tan x}$, $0 < x < \pi$.

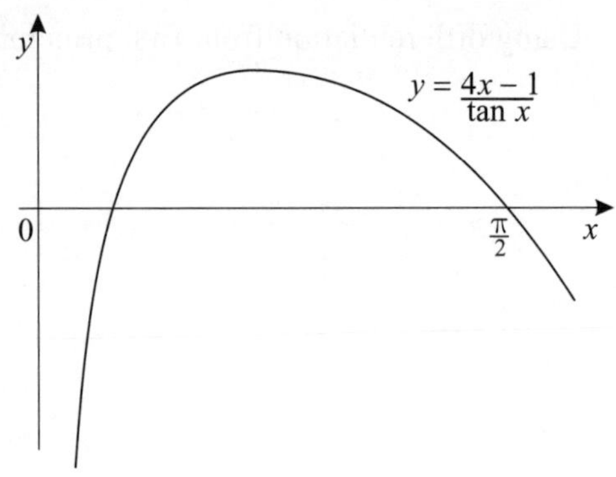

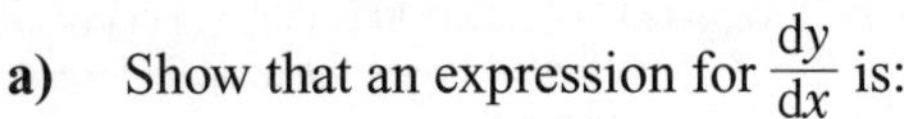

a) Show that an expression for $\dfrac{dy}{dx}$ is:

$$\frac{dy}{dx} = 4\cot x - (4x-1)\,\mathrm{cosec}^2\,x$$

(3 marks)

b) The curve has a maximum in the range $0 < x < \dfrac{\pi}{2}$.
Show that at the maximum point, $2\sin 2x - 4x + 1 = 0$.

(4 marks)

10 A curve is defined by the parametric equations $x = t^2 + 1$, $y = t^3 + 2t$.

Find the equation of the tangent to the curve at the point where $t = 1$.
Give your answer in the form $y = mx + c$.

....

(5 marks)

Differentiation 2

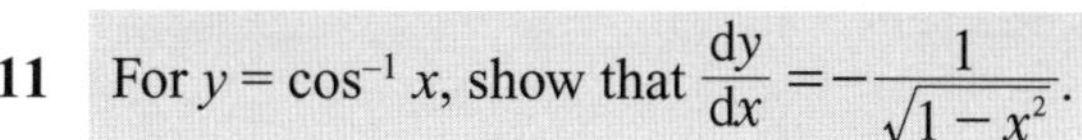

11 For $y = \cos^{-1} x$, show that $\frac{dy}{dx} = -\frac{1}{\sqrt{1-x^2}}$.

(4 marks)

12 The curve C is defined by the parametric equations $x = \frac{\sin\theta}{2} - 3$, $y = 5 - \cos 2\theta$.

a) Find an expression for $\frac{dy}{dx}$ in terms of θ.

....

(2 marks)

b) Hence find the equation of the tangent to C at the point where $\theta = \frac{\pi}{6}$.
Give your answer in the form $ax + by + c = 0$.

....

(3 marks)

c) The line $y = -8x - 20$ crosses C at point P. Find the coordinates of P, showing your working clearly.

....

(6 marks)

d) Find a Cartesian equation for C in the form $y = f(x)$.

....

(3 marks)

Differentiation 2

13 A curve has the equation $x^3 + x^2y = y^2 - 1$.

a) Find an expression for $\frac{dy}{dx}$.

....

(4 marks)

The points P and Q lie on the curve. P has coordinates $(1, a)$ and Q has coordinates $(1, b)$.

b) Find the values of a and b, given that $a > b$.

....

(2 marks)

c) Find the equation of the normal to the curve at Q.
Give your answer in the form $y = mx + c$.

....

(3 marks)

14 A curve has the equation $\sin \pi x - \cos\left(\frac{\pi y}{2}\right) = 0.5$, for $0 \leq x \leq 2$, $0 \leq y \leq 2$.

The curve has one stationary point. Find the coordinates of the stationary point.

....

(7 marks)

Differentiation questions can get quite tricky. First of all, make sure you're using the correct rule, then set out your working clearly and take care when you're simplifying expressions. The actual differentiation is often just the first step in answering a question. So after you've done all that hard work, make sure you interpret your answer correctly to find what you're asked for.

Score

102

Integration 1

Integration starts off pretty easy, but then gets really hard really quickly. Just keep your wits about you, and watch out for places where you can use clever tricks (like when the numerator is the derivative of the denominator).

1 Find $\int \left(\frac{x^2+3}{\sqrt{x}}\right) dx$.

$= (x^2+3)(x^{-1/2})$

$= x^{3/2}$

(3 marks)

2 The curve C has the equation $y = f(x)$, $x > 0$. $f'(x)$ is given as $2x + 5\sqrt{x} + \frac{6}{x^2}$.

A point P on curve C has the coordinates (3, 7). Find $f(x)$, giving your answer in its simplest form.

(6 marks)

3 Region A is bounded by the curve $y = \frac{2}{\sqrt{x^3}}$ $(x > 0)$, the x-axis and the lines $x = 2$ and $x = 4$.

Show that the area of A is $2\sqrt{2} - 2$.

(5 marks)

4 Evaluate the definite integral $\int_p^{4p} \left(\frac{1}{\sqrt{x}} - 4x^3\right) dx$, where $p > 0$, leaving your answer in terms of p.

(4 marks)

Integration 1

5 Find the exact value of p given that $\int_{2p}^{6p} \frac{x^3 + 4x^2}{x^3}\,dx = 4\ln 12$.

(4 marks)

6 Find $\int \frac{1}{x(3x-2)}\,dx$.

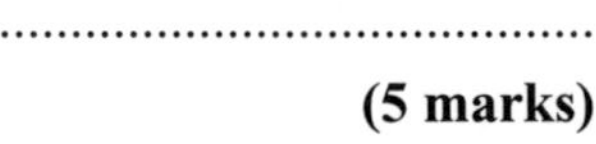

(5 marks)

7 $f(x) = \frac{3x+5}{(3x+1)(1-x)}$

a) Write $f(x)$ in partial fractions.

....

(3 marks)

b) Find the exact value of $\int_{-2}^{4} f(x)\,dx$ in the form $\ln \frac{p}{q}$, where p and q are integers.

....

(3 marks)

8 Find the exact value of $\int_{-1}^{1} \frac{4x-10}{4x^2+4x-3}\,dx$, giving your answer in the form $\ln k$, where k is an integer.

....

(7 marks)

Integration 1

9 Find the exact value of $\int_{\frac{\pi}{12}}^{\frac{\pi}{8}} \sin 2x \, dx$.

....

(3 marks)

10 Find $\int \left(3\tan^2\left(\frac{x}{2}\right) + 3\right) dx$.

....

(3 marks)

11 Find $\int \sec^2 x \, e^{\tan x} \, dx$.

....

(2 marks)

12 The graph below shows part of a curve with equation $y = \frac{2}{3(\sqrt[3]{5x-2})}$.

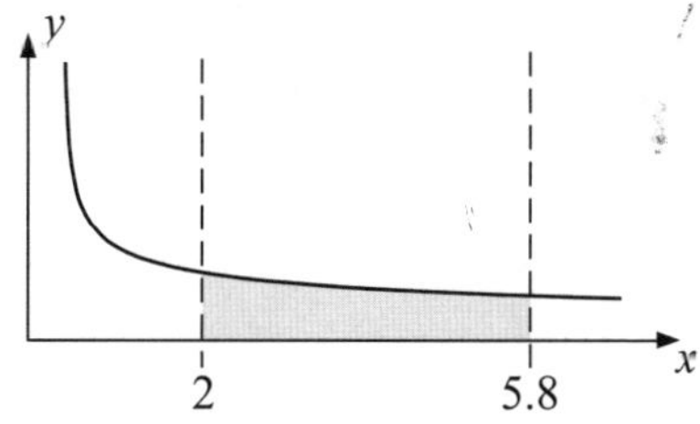

Find the area of the shaded region bounded by the curve, the x-axis, and the lines $x = 2$ and $x = 5.8$.

....

(4 marks)

Integration 1

13 Find the possible values of k that satisfy $\int_{\sqrt{2}}^{2} (8x^3 - 2kx)\, dx = 2k^2$, where k is a constant.

....

(5 marks)

14 The curve $y = 2x^3 - 3x^2 - 11x + 6$ is shown below. It crosses the x-axis at $(-2, 0)$, $(0.5, 0)$ and $(3, 0)$.

Find the area of the shaded region bounded by the curve, the x-axis and the lines $x = -1$ and $x = 2$.

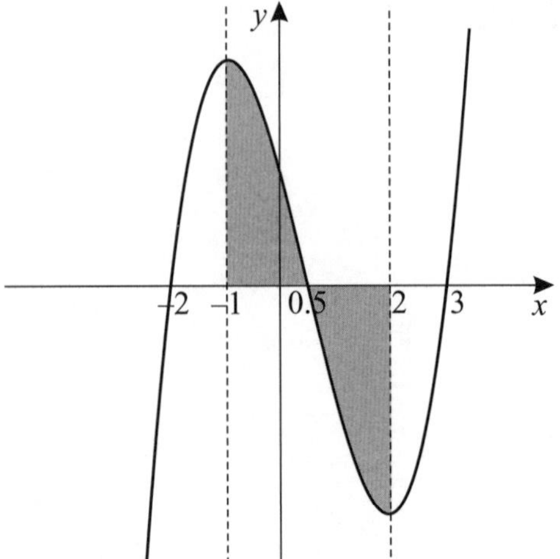

....

(6 marks)

15 The graph on the right shows the curves $y = \frac{8}{x^2}$ and $y = 9 - x^2$ for $x, y \geq 0$.

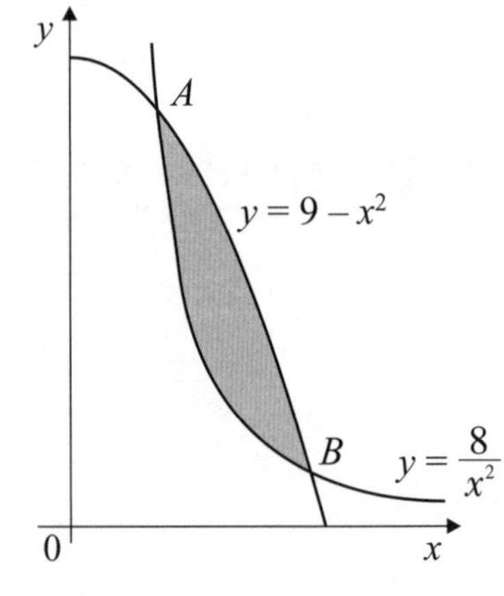

a) Find the coordinates of points A and B.

....

(5 marks)

b) Find the shaded area. Give your answer in the form $a + b\sqrt{2}$, where a and b are fractions.

....

(4 marks)

Integration 1

16 The graph on the right shows the curve C, which has equation $y = \sqrt{x} - \frac{1}{2}x^2 + 1$ $(x \geq 0)$.

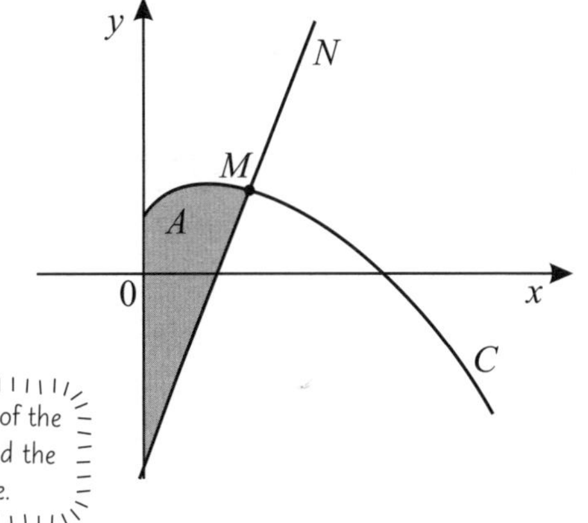

Point M lies on the curve and has coordinates $\left(1, \frac{3}{2}\right)$, and line N is the normal to the curve at point M. The shaded region A is bounded by the y-axis, the curve and the normal to the curve at M.

Find the area of A.

Combine the equations of the curve and the line to find the function to integrate.

....

(9 marks)

17 Jo wishes to find the area (A) enclosed by the graph with equation $y = x^2 + 1$, the line $y = 2$ and the y-axis.

a) Show that the required area is given by $A = \int_1^2 \sqrt{y-1}\,\mathrm{d}y$.

(2 marks)

b) Hence find the value of A.

....

(3 marks)

EXAM TIP

Make sure you're completely happy with integrating powers — including the ones that give you a natural log (ln) when you integrate. It's usually best to write roots and fractions as powers of x — it makes it much easier to integrate (and means you're less likely to make a mistake). To find tricky areas, try splitting them into different bits and finding each bit separately.

Score

86

Integration 2

As is typical with Hollywood blockbusters, after the box office success of Integration 1, here comes the sequel — Integration 2: Integration With A Vengeance. Just remember, in calculus, no one can hear you scream...

1 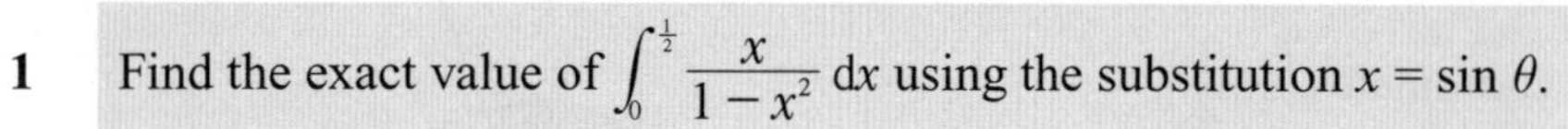
Find the exact value of $\int_0^{\frac{1}{2}} \frac{x}{1-x^2}\,dx$ using the substitution $x = \sin\theta$.

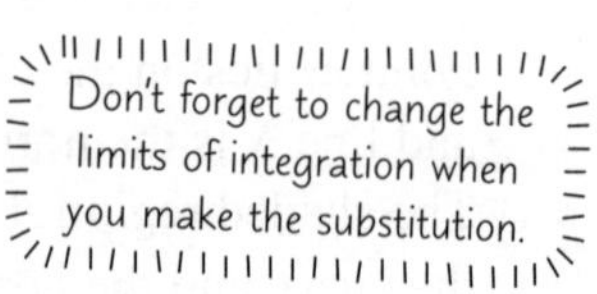

....

(7 marks)

2 Find $\int_1^2 \left(\frac{\ln x}{\sqrt{x}}\right)^2 dx$, using the substitution $u = \ln x$. Give your answer to 3 significant figures.

....

(5 marks)

3 Find $\int 4xe^{-2x}\,dx$.

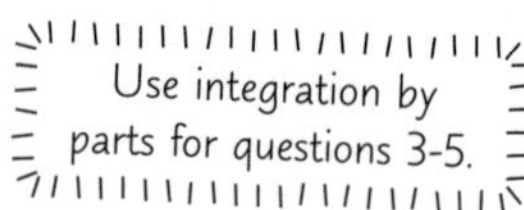

....

(4 marks)

Integration 2

4 Find:

a) $\int x \sin 4x \, \mathrm{d}x$

....

(4 marks)

b) $\int x^2 \cos 4x \, \mathrm{d}x$

....

(4 marks)

5 Calculate the exact value of $\int_1^4 \frac{\ln x}{2x^2} \, \mathrm{d}x$.

....

(6 marks)

Integration 2

6 An ecologist is monitoring the population of newts in a colony. The rate of increase of the population is directly proportional to the square root of the current number of newts in the colony. When there were 36 newts in the colony, the rate of change was calculated to be 0.36.

a) Formulate a differential equation to model the rate of change, in terms of the variables N (number of newts), t (time in weeks).

....

(4 marks)

b) After more research, the ecologist decides that the differential equation $\frac{dN}{dt} = \frac{kN}{\sqrt{t}}$, for a positive constant k, is a better model for the population.
When the ecologist began the survey, the initial population of newts in the colony was 25.

(i) Solve the differential equation, leaving your answer in terms of k and t.

....

(3 marks)

(ii) Given that the value of k is 0.05, calculate how long (to the nearest week) it will take for the population to double.

....

(3 marks)

7 A supermarket sets up an advertising campaign to increase sales on the cheese counter. After the start of the campaign, the number of kilograms of cheese sold each day, S, increases over time, t days. The increase in sales is modelled by the differential equation $\frac{dS}{dt} = k\sqrt{S}$ $(k > 0)$.

a) At the start of the campaign, the supermarket was selling 81 kg of cheese a day.
Use this information to solve the differential equation, giving S in terms of k and t.

....

(3 marks)

b) Given that $\frac{dS}{dt} = 18$ at the start of the campaign, calculate the number of kg sold on the fifth day after the start of the campaign ($t = 5$).

....

(3 marks)

c) How many days will it take before the sales reach 225 kg a day?

....

(2 marks)

Integration 2

8 A solid hemisphere of radius r cm and surface area S cm^2 is decreasing in size.

a) The rate of decrease of r, over time t minutes, is directly proportional to rt.

(i) Formulate a differential equation in terms of r, t and a positive constant k.

....

(2 marks)

(ii) Show that $\frac{dS}{dt} = -2ktS$.

The surface area of a sphere of radius r is $4\pi r^2$.

(4 marks)

b) The total surface area of the hemisphere is 200 cm^2 at time $t = 10$ minutes and 50 cm^2 at time $t = 30$ mins.

(i) Find the particular solution of the equation $\frac{dS}{dt} = -2ktS$.
Give the values of any constants to 3 significant figures.

....

(5 marks)

(ii) Hence find the initial surface area of the hemisphere to 3 significant figures.

....

(1 mark)

c) A solid sphere with the same initial radius as the hemisphere is also decreasing in size. Maddy decides to use the differential equation given in part a) (ii) to model the surface area of the sphere. Explain why this will not be appropriate.

....

....

(1 mark)

Does your brain hurt? Mine does. You're given the formula for integration by parts in the exam paper, so you can always look it up if you're struggling to remember it — but make sure you know exactly how to use it. Questions on differential equations can look pretty nasty, but once you've separated the variables, it shouldn't be too hard to integrate each side.

Score

61

Numerical Methods

There are two main parts to numerical methods — finding roots using iteration (including my personal favourite, the Newton-Raphson method) and numerical integration (which is also pretty darn exciting if you ask me).

1 The graph below shows the function $f(x) = 4(x^2 - 1)$, $x \geq 0$, and its inverse function $f^{-1}(x)$.

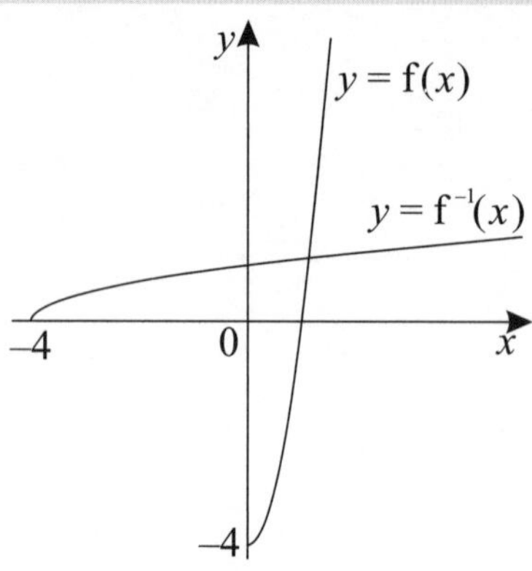

a) By finding an expression for $f^{-1}(x)$ and considering how the graphs are related, show that $\sqrt{\frac{x}{4} + 1} - x = 0$ at the point where the graphs meet.

(4 marks)

b) Show that the equation $\sqrt{\frac{x}{4} + 1} - x = 0$ has a root in the interval $1 < x < 2$.

(2 marks)

c) Starting with $x_0 = 1$, use the iteration formula:

$$x_{n+1} = \sqrt{\frac{x_n}{4} + 1}$$

to find the x-coordinate of the point of intersection, correct to 3 significant figures.

....

(3 marks)

d) Does the equation from part a) have any other roots? Explain your answer.

Look at the graph above...

....

....

(1 mark)

Numerical Methods

2 The sketch below shows the intersection of the curve $y = 6^x$ with the line $y = x + 2$ at the point P.

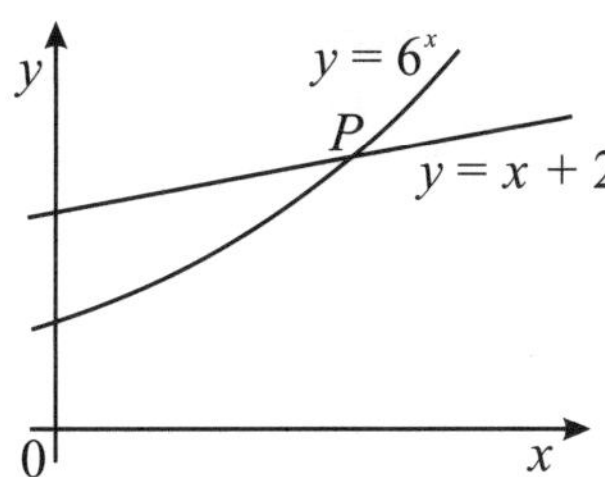

a) Show that a root of the equation $6^x - x - 2 = 0$ lies in the interval [0.5, 1].

(2 marks)

b) Show that the Newton-Raphson iteration formula for finding the x-coordinate of P can be written as:

$$x_{n+1} = \frac{6^{x_n}(x_n \ln 6 - 1) + 2}{6^{x_n} \ln 6 - 1}$$

Remember — the derivative of a^x is $a^x \ln a$.

(4 marks)

c) Using the Newton-Raphson formula with a starting value of $x_0 = 0.5$, find the x-coordinate of P correct to 4 significant figures.

....

(3 marks)

d) By considering an appropriate interval, verify that this value is accurate to 4 significant figures.

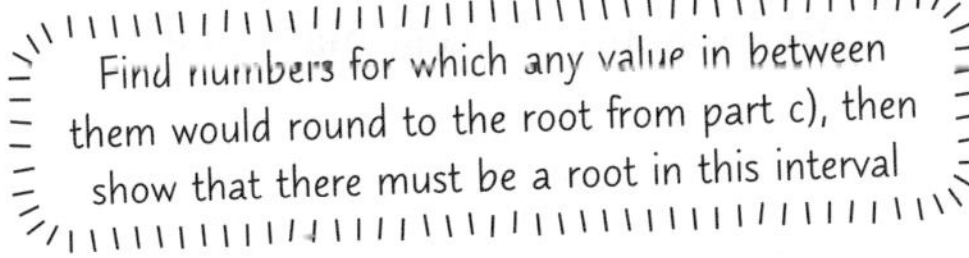

(3 marks)

e) k is chosen as a different starting value. The gradient of the tangent at $x = k$ is 0. Explain why the Newton-Raphson method will fail for this starting value.

....

....

(1 mark)

Numerical Methods

3 $f(x) = x^3 - x^2 + 4$

a) The equation $f(x) = 0$ has a root in the interval $(-2, -1)$.
For the starting value $x_0 = -1.5$, $f(x_0) = -1.625$ and $f'(x_0) = 9.75$.
Use the Newton-Raphson method to obtain a value for x_1, the second approximation for the root.
Give your answer to 4 significant figures.

....

(2 marks)

b) Explain why the Newton-Raphson method fails when $x_0 = \frac{2}{3}$.

....

....

(2 marks)

4 The diagram on the right shows the graph of $y = 2^{x^2}$.

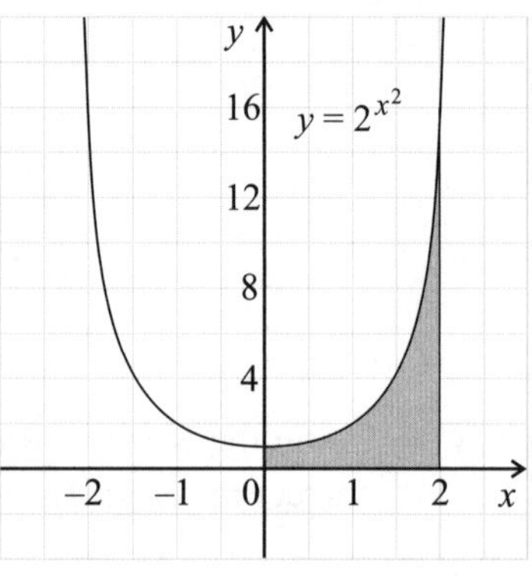

a) Use the trapezium rule with 4 strips to find an estimate for the area of the region bounded by the axes, the curve and the line $x = 2$.
Give your answer to 3 significant figures.

....

(4 marks)

b) Given that the curve is concave upwards for all values of x, explain whether the estimate in a) is an overestimate or an underestimate.

....

....

(1 mark)

c) Suggest one way to find a more accurate estimate for this area using the trapezium rule.

....

....

(1 mark)

Numerical Methods

5 Figure 1 shows the graph of $y = x \sin x$. The region R is bounded by the curve and the x-axis ($0 \leq x \leq \pi$).

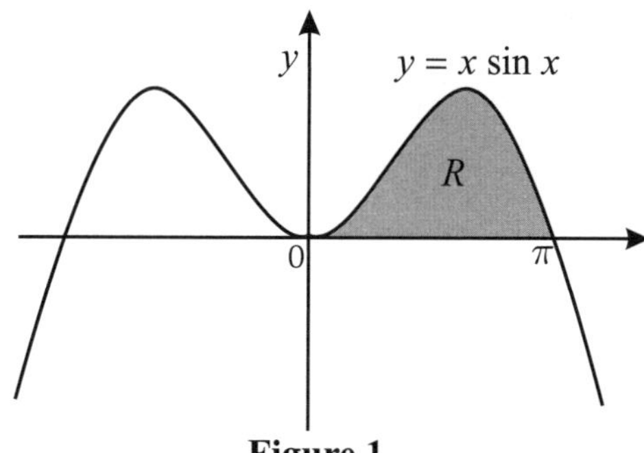

Figure 1

a) Fill in the missing values of y in the table below.
Give your answers to 5 significant figures.

x	0	$\frac{\pi}{4}$	$\frac{\pi}{2}$	$\frac{3\pi}{4}$	π
y	0	0.55536			0

(1 mark)

b) Hence find an approximation for the area of R, using the trapezium rule and all the values in the table.
Give your answer to 4 significant figures.

....

(3 marks)

c) Find the exact area of R using integration by parts.

....

(6 marks)

d) Hence find the percentage error of the approximation found in part b).
Give your answer to 2 significant figures.

Leave the answer from part b) in your calculator to get a more accurate value.

....

(2 marks)

Numerical Methods

6 Use the trapezium rule with 5 strips to estimate $\int_{1.5}^{4} (3x - \sqrt{2^x})\, dx$.
Give your answer to 4 significant figures.

....

(4 marks)

7 The graph below shows the curve $y = \frac{3\ln x}{x^2}$, $x > 0$.
The shaded region R is bounded by the curve, the line $x = 2.5$ and the line $x = 3$.

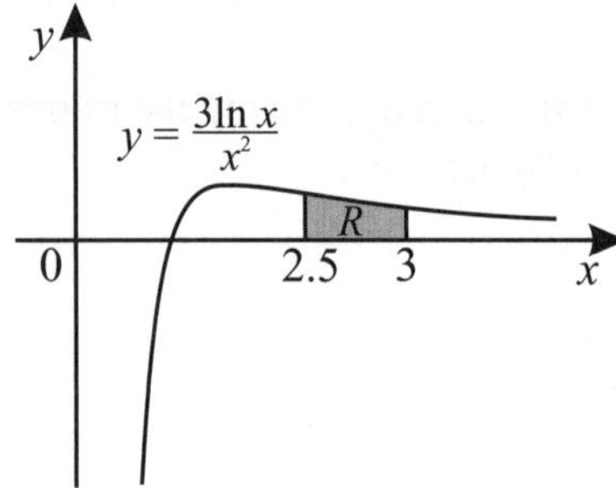

a) Find an approximation for the area of R, using the trapezium rule with 5 strips.
Give your answer to 4 significant figures.

....

(4 marks)

b) By considering the areas of appropriate rectangles, show that the area of R is 0.2, correct to 1 decimal place.

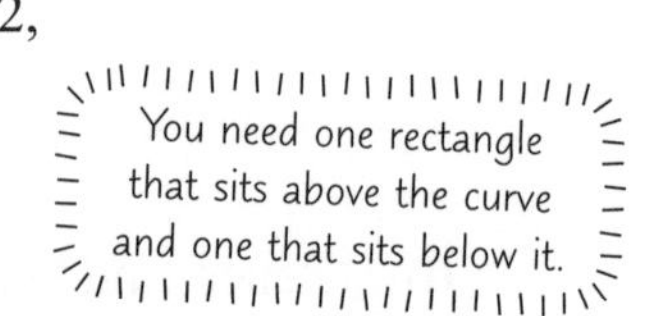

(3 marks)

A bit of calculator wizardry can save a lot of time in the exam — especially for iteration questions. Put in your starting value of x and press =, then input the rest of your iteration equation (or Newton-Raphson formula) in terms of ANS. Each time you press =, you'll get the next value of x (without having to type out the full formula each time). What a lifesaver.

Score

56

Vectors

Aaah, good old dependable vectors. They've got a certain magnitude about them, and they always have a clear direction. Much like an inspirational leader. It's kind of how I see myself after I've brought about the revolution.

1 A vector of magnitude 7 acts vertically upwards. Another vector, of magnitude $4\sqrt{2}$, acts at an angle of 45° below the positive horizontal direction.
The resultant of the two vectors is **r**.

a) Find **r** in terms of **i** and **j**.

r = ...

(2 marks)

b) The vector **s** acts parallel to **r**, and has magnitude 35.
Find **s** in terms of **i** and **j**.

s = ...

(3 marks)

2 Points A, B and C have position vectors $-\mathbf{i} + 7\mathbf{j} - 2\mathbf{k}$, $5\mathbf{i} - 3\mathbf{j} + 6\mathbf{k}$ and $5\mathbf{i} + 4\mathbf{j} + 3\mathbf{k}$ respectively.
M is the midpoint of AB.

Find the exact value of k such that $|\overrightarrow{CM}| = k|\overrightarrow{AB}|$.

k = ...

(5 marks)

3 Given that $\overrightarrow{OA} = -2\mathbf{i} + 4\mathbf{j} - 5\mathbf{k}$, $\overrightarrow{OB} = 14\mathbf{i} + 12\mathbf{j} - 9\mathbf{k}$ and $\overrightarrow{OC} = 2\mathbf{i} + \mu\mathbf{j} + \lambda\mathbf{k}$, find the values of μ and λ such that points A, B and C are collinear.

If two vectors are parallel and share a point, they lie on the same straight line.

μ = ... λ = ...

(5 marks)

Vectors

4 Points A, B and C have position vectors $\begin{pmatrix} 1 \\ -3 \\ 2 \end{pmatrix}$, $\begin{pmatrix} 4 \\ -12 \\ 8 \end{pmatrix}$ and $\begin{pmatrix} -3 \\ 9 \\ -6 \end{pmatrix}$ respectively.

Point D lies on AB such that $AD:DB = 2:1$.

Point E is positioned such that $\overrightarrow{OD} = -\frac{1}{2}\overrightarrow{CE}$.

Find the position vector of point E.

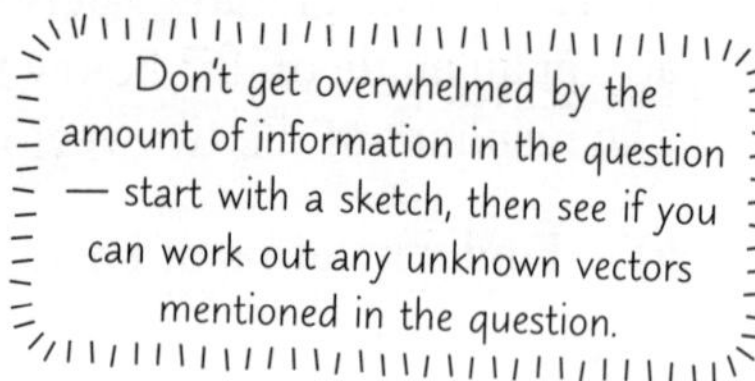

....

(5 marks)

5 Figure 1 shows a sketch of a triangle, PQR. Given that $\overrightarrow{PQ} = \begin{pmatrix} 2 \\ -9 \\ 3 \end{pmatrix}$ and $\overrightarrow{QR} = \begin{pmatrix} 14 \\ 6 \\ 7 \end{pmatrix}$, find the angle $\angle QPR$. Give your answer in degrees to 1 decimal place.

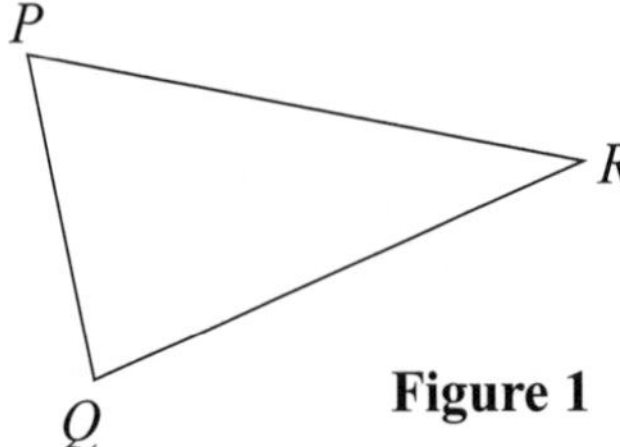

Figure 1

.... °

(5 marks)

Vectors

6 Two drones, A and B, take off from a launch pad at origin O and follow a series of movements based on the vectors $\mathbf{a} = (4\mathbf{i} + 6\mathbf{j} + 5\mathbf{k})$ metres, $\mathbf{b} = (-\mathbf{i} - 2\mathbf{j} - 2\mathbf{k})$ metres and $\mathbf{c} = (-3\mathbf{j} + \mathbf{k})$ metres, where $\mathbf{i}$ and $\mathbf{j}$ are horizontal unit vectors and $\mathbf{k}$ is a vertical unit vector with the upwards direction being positive.

Drone A follows the vectors $\mathbf{a}$, then $\mathbf{b}$, then $\mathbf{c}$, and then maintains its final position.
Drone B follows the vectors $2\mathbf{a}$, then $\mathbf{b}$, then $-3\mathbf{c}$, and then maintains its final position.

a) Calculate to 3 s.f. the distance between the final positions of drones A and B.

.... m

(4 marks)

b) Drone A now moves in a straight line to a position exactly 2 m directly below drone B. Find the vector that describes this movement of drone A.

....

(2 marks)

c) To maintain contact with the controller on the ground, drone B must always stay within 50 m of the origin O. How far can drone B move in the positive $\mathbf{j}$ direction from its current position before it passes out of range? Give your answer to 3 s.f.

.... m

(5 marks)

Fun, fun, fun, vectors are just pure fun, don't you think... ... ... Well anyway, make sure you're comfortable finding their magnitudes and using trig to find angles. Don't be put off by 3D ones — they're generally pretty much the same as their 2D counterparts. Some questions can be quite involved, but always start with a sketch, take it step by step and you'll be OK.

Score

36

Data Presentation and Interpretation

As I am sure you will agree, the only thing more exciting than presenting data is interpreting it. This is exactly the kind of topic where the examiners will test your large data set skills, so take another look at those spreadsheets.

1 Penny wants to find out about the scores achieved by teams taking part in a pub quiz. She takes a random sample of 10 scores (x) and calculates the statistics below:

$$\sum x_i = 500 \text{ and } \sum x_i^2 = 25\,622.$$

a) Find the sample mean and the sample standard deviation for the data.

sample mean = ..., sample standard deviation = ...

(2 marks)

b) Penny decides to include an extra randomly-chosen score in her sample. This new score is 50. Giving reasons, but without further calculation, explain the effect of adding this score on:

(i) the sample mean,

....

....

(2 marks)

(ii) the sample standard deviation.

....

....

(2 marks)

2 The areas (A, in thousands of square kilometres) of a sample of countries are summarised in the table below.

Area (A, thousands of km^2)	$0 \le A < 100$	$100 \le A < 500$	$500 \le A < 1000$	$1000 \le A < 2000$	$2000 \le A < 5000$
Frequency	50	24	15	9	2

Morwenna draws a histogram to represent the data.
The bar for the $0 \le A < 100$ class has a width of 0.5 cm and a height of 20 cm.

Find the width and height of the bar for the $500 \le A < 1000$ class.

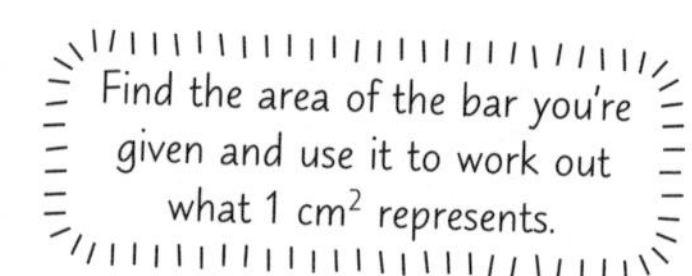

width = ... cm, height = ... cm

(3 marks)

Data Presentation and Interpretation

3 The graph below shows some data about the times taken for runners to finish the Alverston marathon. The graph shows the mean time taken, in minutes, by different age groups of runners who ran the marathon in 2000 and 2010.

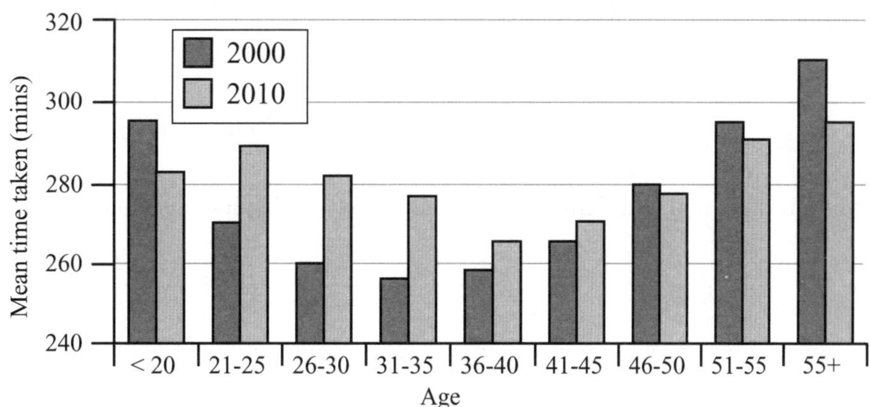

a) A researcher studies a randomly-chosen sample of runners who were in the 21-25 age group in 2000. Does the graph suggest these runners' times improved in 2010? Justify your answer.

....

....

....

(2 marks)

b) Write down one assumption you have made about the runners in your answer to part a).

....

....

(1 mark)

4 Naga has summarised some data from a large data set, showing the mean weight of people in different marital-status categories. Naga's summary is shown below.

Marital Status	Married	Widowed	Divorced	Never married	Living with partner	Separated
Mean weight (kg)	78.4	67.7	84.4	74.9	82.7	85.8

Naga suggests finding the mean weight of all the above people using the calculation shown below. Explain why this is not a suitable method for this data set.

$$\frac{78.4 + 67.7 + 84.4 + 74.9 + 82.7 + 85.8}{6} = 79$$

....

....

....

(2 marks)

Data Presentation and Interpretation

5 The sales figures for a gift shop over a 12-week period are shown on the stem and leaf diagram below.

Stem	Leaf
3	8
4	1 2 6 9
5	5 8 9
6	0 2 4
7	
8	
9	1

Key
3 | 8 means £3800

a) Explain why the stem and leaf diagram is a suitable way to display this data.

....

....

(1 mark)

b) Find the median, lower quartile, upper quartile and interquartile range of the sales data.

median = £ ...., lower quartile = £ ....,

upper quartile = £ ...., interquartile range = £ ....

(4 marks)

The shop's manager is considering excluding any outliers from his analysis of the data, to get a more realistic idea of how the shop is performing.

c) Using a calculation, explain why the data value £9100 is an outlier.

....

....

(1 mark)

d) Do you think the manager should include this outlier in his analysis? Explain your answer.

....

....

(1 mark)

e) The mean and standard deviation for this sales data are calculated to be: mean = £5540, standard deviation = £1370 (both to 3 s.f.). Do you think these two measures, or the median and interquartile range, are more useful measures of location and spread for this data? Explain your answer.

....

....

....

(2 marks)

Data Presentation and Interpretation

6 A sample of Year 12 students in a school were asked how much their monthly phone bill is (in £). The incomplete table and histogram below show the results for the sample.

a) Complete the table and the histogram.

Monthly phone bill, £b	Frequency
$0 \leq b < 10$	12
$10 \leq b < 15$	23
$15 \leq b < 18$	
$18 \leq b < 20$	
$20 \leq b < 25$	18
$25 \leq b < 35$	6

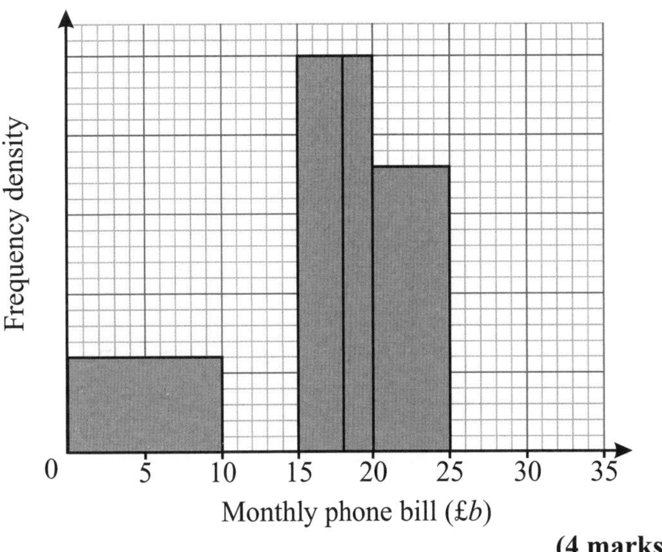

(4 marks)

b) Estimate the proportion of students in the sample who have a monthly phone bill of between £12.50 and £17.50. Give your answer to 3 significant figures.

....

(2 marks)

c) Estimate the sample mean and sample standard deviation for the data above.

Use your calculator functions to estimate these values.

Sample mean = £ ...., Sample standard deviation = £ ....

(3 marks)

d) Estimate the 20th percentile of the sample data.

£ ....

(3 marks)

Data Presentation and Interpretation

7 The median house price (in £) in two different areas in the UK for each year from 2004 to 2015 is shown on the line graph below.

- For Barking and Dagenham, the mean of these prices is approximately £181 800 and the standard deviation is £23 776.
- For the South West of England, the mean is approximately £186 000 and the standard deviation is £12 808.

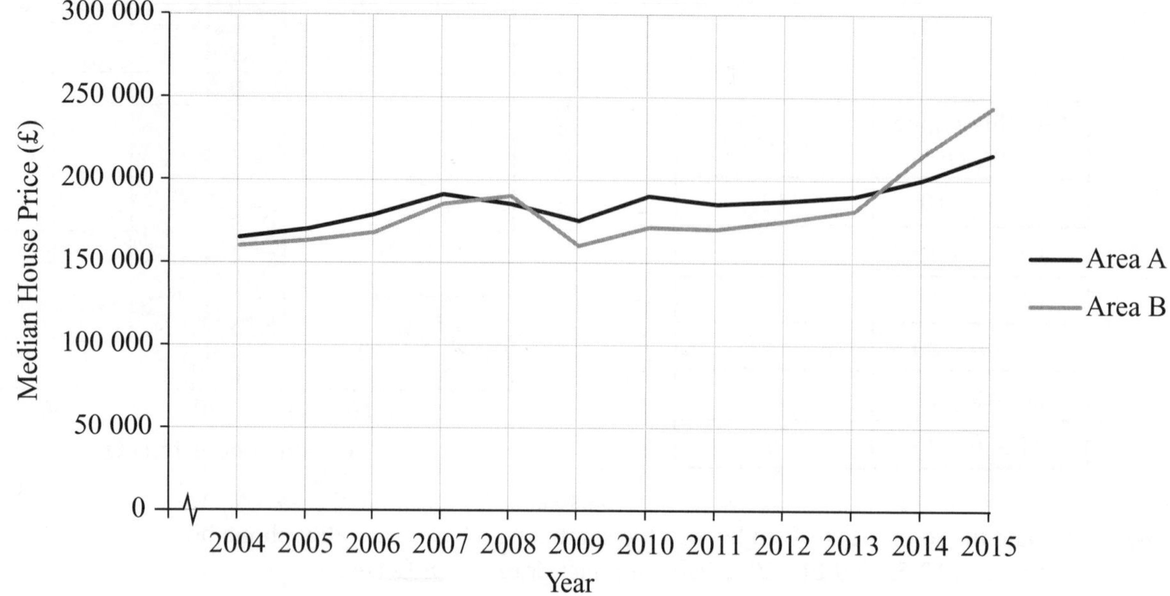

a) Match the areas Barking and Dagenham and the South West of England to the corresponding line on the diagram. You must explain your reasoning.

Area A = ...

Area B = ...

Reasoning: ....

....

....

(2 marks)

For Area A, the lowest yearly median house price is £165 000 and the highest is £215 000.
For Area B, the lowest yearly median house price is £160 000 and the highest is £243 500.

b) Using the information given above, show that the data sets for areas A and B both contain at least one outlier. You must show your working.

(3 marks)

Data Presentation and Interpretation

8 The heights of giraffes living in a zoo were measured.
A cumulative frequency diagram of the results is shown below.

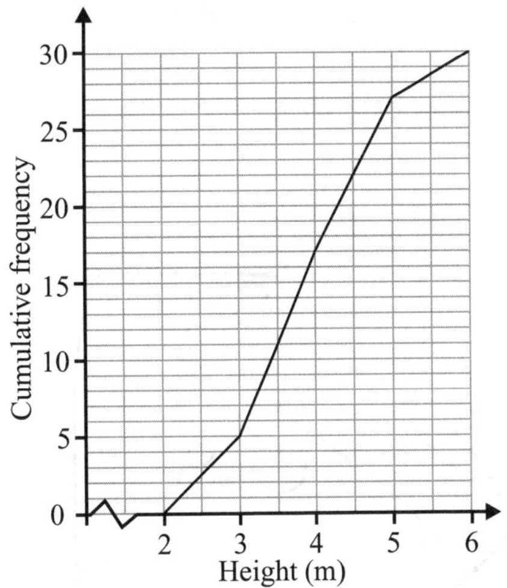

a) Use the diagram to estimate the 60th percentile of the data.

.... m

(1 mark)

The data about heights of the giraffes in the zoo is summarised in the box plot below.

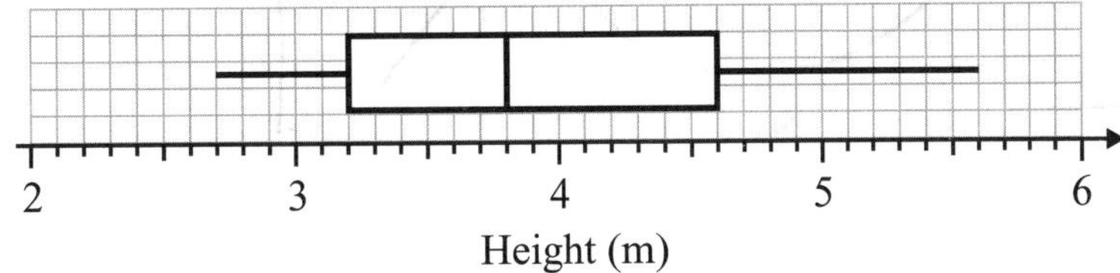

The heights of giraffes living in a game reserve were also measured.
This data is summarised in the box plot below.

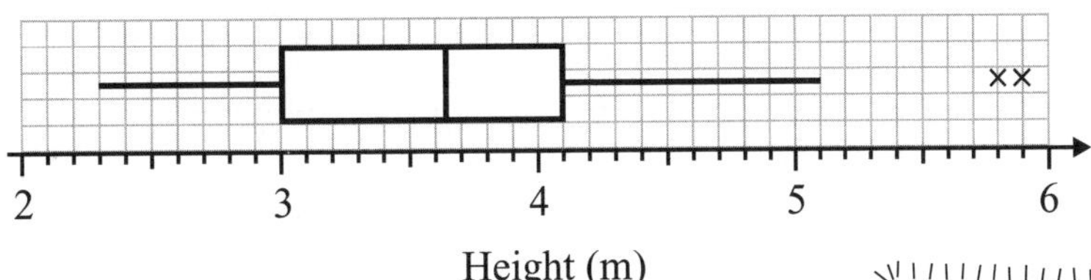

You should talk about the shape/skewness of the distributions in your answer.

b) Compare the heights of the two groups of giraffes.

....

....

....

....

....

(3 marks)

As well as being able to calculate measures like the mean and the standard deviation, it's important to know what they tell you about a data set. Once you understand what they are and why they're useful, questions asking you to interpret or discuss data become a breeze — which means picking up those extra marks in the exam without needing to do any tricky maths.

Score

44

Probability

Venn diagrams. My question is, Vhat about the Vhere, the Vhich and the Vhy diagrams? I'm sorry, that isn't even remotely funny. In fact, I often find that me making jokes and people laughing are mutually exclusive events...

1 The events A and B are mutually exclusive. P(A) = 0.1 and P(B) = 0.4. Event C has probability P(C) = 0.3.

Events B and C are independent, and the probability of both events A and C occurring is 0.06.

a) Draw a Venn diagram showing the probabilities of events A, B and C.

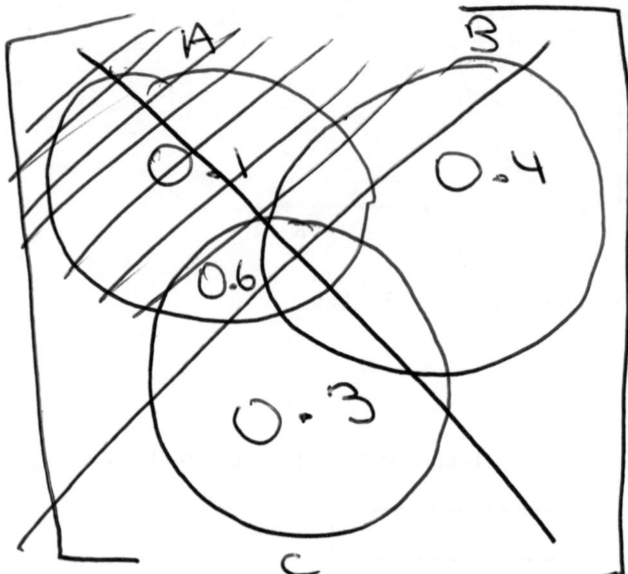

(5 marks)

b) Are events A and C independent? Explain your answer.

....

....

(2 marks)

c) Find P(B ∪ C)

....

(1 mark)

d) Find P(A' ∩ B')

....

(1 mark)

e) Find P(B' | A'), to 2 decimal places.

....

(2 marks)

Probability

2 This incomplete Venn diagram shows the probabilities of two independent events L and M. Calculate P(L' ∩ M').

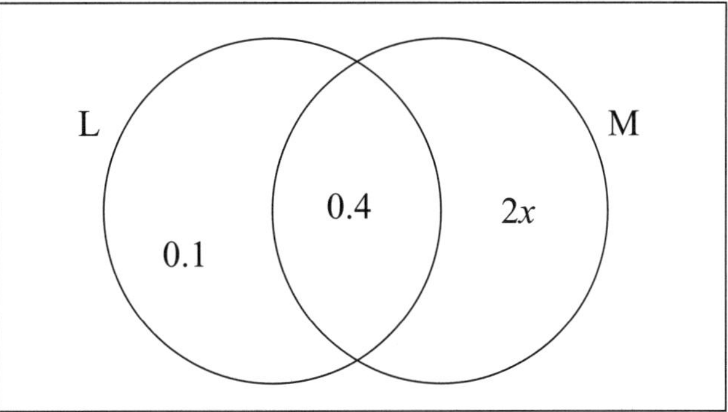

....

(4 marks)

3 Of 30 drivers interviewed, 9 have been involved in a car crash in the last year. Of those who have been involved in a crash, 5 wear glasses. The probability of wearing glasses, given that the driver has not been involved in a car crash is $\frac{1}{3}$.

Find the probability that a glasses-wearer interviewed has been involved in a car crash.

....

(4 marks)

Probability

4 Alan is studying the age of people attending Ulverston Dynamo's home matches. He surveyed every fan attending the match against Cark Sportif, and recorded their ages and gender in the table shown below.

Age Group	Frequency	
	Male	Female
Under 18	305	207
18-30	431	333
31-45	378	531
46-60	472	236
Over 60	359	115

a) Find the probability that a fan selected at random from the crowd is under 18 given that they are male.

....

(2 marks)

b) Alan uses his data to model probabilities for the crowd at Ulverston Dynamo's next home match, which is against AC Flookburgh. He says the probability that a fan selected at random from the crowd is in the 18-30 age group is $\frac{764}{3367}$. Describe one assumption Alan has made in his model, and explain why this assumption may not be valid.

....

....

(2 marks)

5 Jessica has two packs of cards. One pack has some cards missing. The probability of selecting a heart from this pack, P(S) = 0.3. Jessica selects one card from this pack and one card from a complete pack of 52 cards.

a) Find the probability that at least one of the cards is a heart.

....

(2 marks)

Jessica then draws three cards from another complete pack without replacement.

b) Find the probability that she draws at least two hearts, given that the first card she drew was a heart. Give your answer to 3 significant figures.

Without replacement means there'll be a different number of cards each time she draws one — and don't forget about the conditional part too.

....

(3 marks)

c) Give one assumption you made in answering parts a) and b).

....

....

(1 mark)

Probability

6 $P(A \mid B) = 0.31$, $P(B \mid A) = 0.25$ and $P(B) = 0.4$. Find $P(A')$.

....

(3 marks)

7 At a school cafeteria, every child takes at least one type of fruit with their lunch. The cafeteria has bananas, apples and oranges.

65 children use the cafeteria.
25 take a banana, 28 take an orange, 32 take an apple, 5 take a banana and an orange, 6 take an orange and an apple, and 4 take an orange, an apple and a banana.

a) Find the number of children who take a banana and an apple.

....

(4 marks)

Two children are picked at random, one after the other, without replacement.

b) Given that at least one child takes an apple, find the exact probability that the first child took an apple.

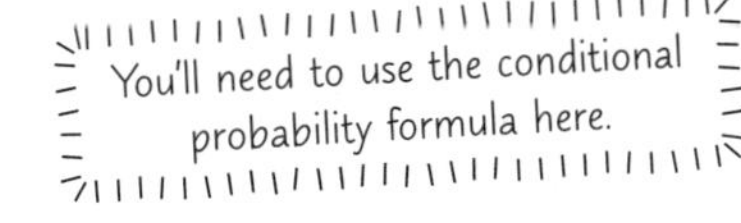

....

(4 marks)

Even if the question doesn't ask you to draw a Venn or tree diagram, it's often helpful to draw one. They make your life so much easier. Also, make sure you understand all the different bits of set notation — it'd be silly to get a question wrong because you mixed up the union (∪) and intersection (∩) symbols. Just remember, 'u for **u**nion and n for i**n**tersection'.

Score

40

Statistical Distributions

Another treat for probability fans (like me), and your first glimpse of the binomial and normal distributions. If you don't warm to these distributions here, you'll get another chance with hypothesis testing in the next topic.

1 The discrete random variable X has the probability function shown below.

$$P(X = x) = \begin{cases} \frac{kx}{6} & \text{for } x = 1, 2, 3 \\ \frac{k(7-x)}{6} & \text{for } x = 4, 5, 6 \\ 0 & \text{otherwise} \end{cases}$$

a) Find the value of k.

$k =$ ....

(2 marks)

b) Find $P(1 < X \leq 4)$.

....

(2 marks)

A discrete random variable Y has the probability function $P(Y = y) = 0.2$ for $y = 1, 2, 3, 4, 5$.

c) State the name of the distribution of Y.

....

(1 mark)

2 The number of points awarded to each contestant in a talent competition is modelled by the discrete random variable X with the following probability distribution:

x	0	1	2	3
$P(X = x)$	0.4	0.3	a	b

A contestant is twice as likely to be awarded 2 points as they are to be awarded 3 points.

By finding the values of a and b, calculate the probability that for two randomly chosen contestants, one scores 2 points and the other scores 3 points.

....

(4 marks)

Statistical Distributions

3 In a game, a player tosses three fair coins. If three heads occur then the player wins 20p. If two heads occur then the player wins 10p. For any other outcome, the player wins nothing.

a) If X is the random variable 'amount won in pence', draw a table to show the probability distribution of X.

(3 marks)

b) The player pays 10p to play each game. Find the probability that the player wins 40p in total over two games, given that they make a profit over the two games.

....

(4 marks)

4 5% of chocolate bars made by a particular manufacturer contain a 'golden ticket'. A student buys 5 of the chocolate bars every week for 8 weeks.

The number of golden tickets he finds is represented by the random variable X.

a) State two assumptions needed for X to follow the binomial distribution B(40, 0.05).

....

....

(2 marks)

Assuming that $X \sim$ B(40, 0.05):

b) Find $P(X > 1)$.

....

(2 marks)

c) How many golden tickets can the student expect to find from the 40 chocolate bars? Justify your answer.

....

(1 mark)

Statistical Distributions

5 A particular model of car, the Dystopia, is prone to developing a rattle in the first year after being made. The probability of any particular Dystopia developing the rattle in its first year is 0.65.

A random sample of 20 one-year-old Dystopias is selected.

a) Find the probability that at least 12 but fewer than 15 of the cars rattle.

....

(3 marks)

b) Find the probability that more than half of the cars rattle.

....

(2 marks)

c) A further five random samples of Dystopias are tested. There are 20 cars in each sample. Find the probability that more than half of the cars in exactly three of these five samples rattle.

Define a new random variable that follows a binomial distribution with $n = 5$ and p = P(more than half of cars rattle).

....

(3 marks)

6 An ice cream shop owner finds that, on 1st July, 880 out of the 1100 customers chose a sugar cone.

a) A random sample of 20 customers from 2nd July is selected. Use a binomial distribution and the data from the previous day to estimate the probability that exactly 12 of them chose a sugar cone.

Use the info to find p.

....

(3 marks)

The owner claims that 42% of customers buy an ice cream with at least one scoop of chocolate ice cream.

b) Assuming that this claim is correct, and that the next 75 customers form a random sample of customers, find the probability that more than 30 of them choose at least one scoop of chocolate ice cream.

....

(3 marks)

c) Comment on the validity of the binomial model you used in part b).

....

....

(1 mark)

Statistical Distributions

7 The histogram below represents the masses of newborn babies in a particular hospital in 2016. The masses have a mean of 3.55 kg and a standard deviation of 0.5 kg.

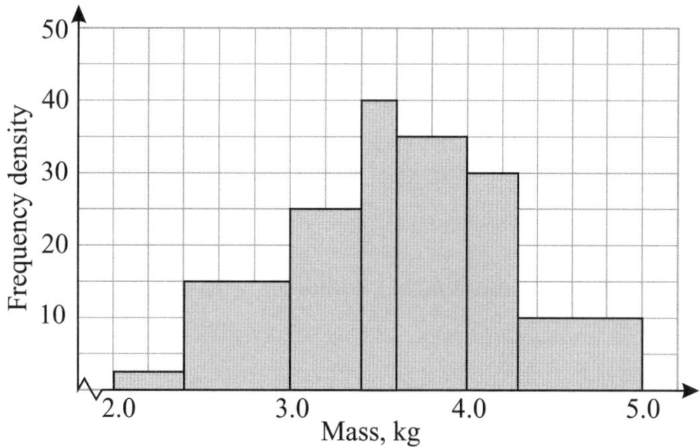

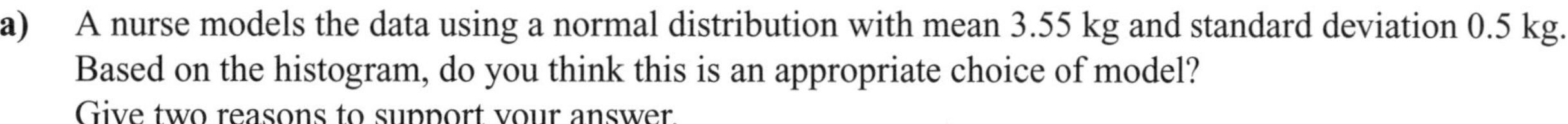

a) A nurse models the data using a normal distribution with mean 3.55 kg and standard deviation 0.5 kg. Based on the histogram, do you think this is an appropriate choice of model? Give two reasons to support your answer.

....

....

(2 marks)

b) Babies who have a mass of less than 2.5 kg at birth are said to have a low birth weight. Using the nurse's model, estimate the probability that a randomly-selected newborn baby at the hospital has a low birth weight.

....

(2 marks)

8 The volumes of water in a certain brand of fire extinguisher approximately follow a normal distribution, as shown in the diagram below. The points of inflection of the graph are labelled.

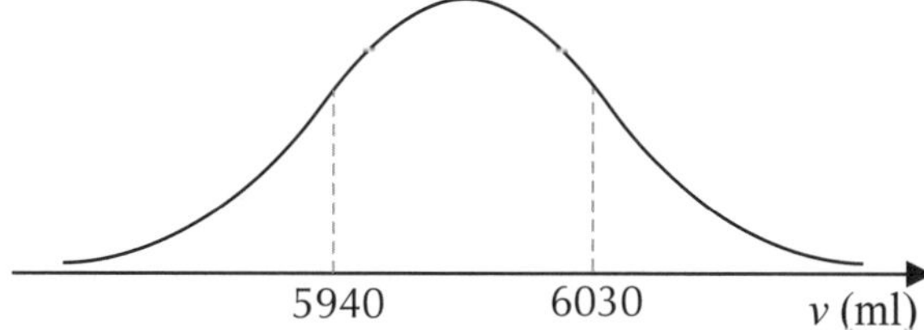

a) Use the diagram to estimate the mean, μ, and standard deviation, σ, for the model.

μ = ... ml, σ = ... ml

(2 marks)

b) Hence use the model to estimate the probability that a randomly-selected fire extinguisher contains between 5900 ml and 6100 ml of water.

....

(2 marks)

Statistical Distributions

9 The exam marks for 1000 candidates can be modelled by a normal distribution with mean 50 marks and standard deviation 15 marks.

a) One candidate is selected at random. Find the probability that they scored fewer than 30 marks on this exam.

....

(2 marks)

b) The pass mark is 41. Estimate the number of candidates who passed the exam.

....

(2 marks)

c) Find, to the nearest whole number, the mark needed for a distinction if the top 10% of the candidates achieved a distinction.

....

(2 marks)

10 The diameters of the pizza bases made at a restaurant are normally distributed. The mean diameter is 12 inches, and 5% of the bases measure more than 13 inches.

a) Find the standard deviation of the diameters of the pizza bases to 3 significant figures.

....

(3 marks)

Any pizza base with a diameter of less than 10.8 inches is considered too small and is discarded.

b) If 100 pizza bases are made in an evening, approximately how many would you expect to be discarded due to being too small?

....

(3 marks)

Three pizza bases are selected at random.

c) Find the probability that at least one of these bases is too small.

....

(3 marks)

Statistical Distributions

11 The time in minutes, X, that it takes a window cleaner to clean each window of an office block is normally distributed with a mean of 4 minutes and a standard deviation of 1.1 minutes.

a) Find the probability that a randomly-selected window takes less than 3.5 minutes to clean.

....

(2 marks)

b) Find the probability that the time taken to clean a randomly-selected window deviates from the mean by more than 1 minute.

....

(3 marks)

c) Find the time taken t, in minutes to 1 decimal place, such that there is a 1% probability that a randomly-selected window will take longer than t minutes to clean.

.... mins

(2 marks)

12 The distances, in miles, run by a group of athletes during one training session can be modelled by a normal distribution with a mean of 8.2 miles and a variance of 0.6 miles.

Distances can be converted from miles to kilometres using the approximation $K \approx 1.6M$, where K is the distance in kilometres and M is the distance in miles.

Find the normal distribution that could be used to model the distances run by these athletes in kilometres.

....

(2 marks)

Statistical Distributions

13 One weekend, 48% of customers at Soutergate Cinema went to see the latest superhero film. Mia selects a random sample of 200 of the customers. She defines X as the number of customers in the sample who went to see the superhero film, and models X using a binomial distribution.

a) Write down the binomial distribution that Mia could use to model X.

....

(1 mark)

Mia plots the probability distribution for X, shown below.

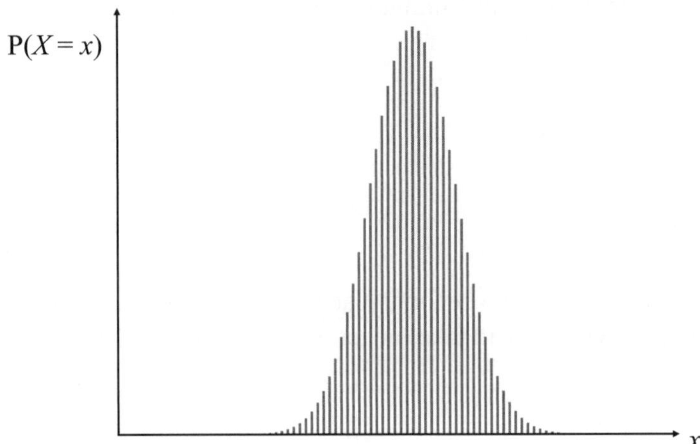

b) With reference to the graph above, explain why it would be suitable for Mia to use a normal distribution to approximate the distribution of X.

....

....

(1 mark)

c) Explain why Mia would need to use a continuity correction on a normal approximation for this distribution.

....

....

(1 mark)

d) Mia uses the mean of the binomial distribution from part a) as the mean of the normal approximation. For the variance of the normal approximation, she uses npq. Find the normal approximation Mia should use.

....

(2 marks)

In the exam, you'll be given a percentage-points table showing a few values from the standard normal distribution. But you'll mainly be using the binomial and normal functions on your calculator to find probabilities — so make sure you know exactly what each one does and when you should use it. And always think about whether a distribution is appropriate for the situation.

Score

73

Statistical Hypothesis Testing

Here we enter the murky world of hypothesis testing, but we welcome back our friends the binomial and normal distributions. You should remember them from the previous topic... they remember you...

1 Josie wants to find out what pupils in her school think about politics. She decides to ask a sample of the pupils.

a) Identify the population that Josie is interested in.

....

(1 mark)

b) Josie plans to select everyone in her A-level politics class as her sample. Name Josie's sampling method and explain whether or not her sample is likely to be representative of the population.

....

....

(2 marks)

c) Josie's friend Mike suggests she randomly selects 10 pupils from each year and asks them to send their responses to her questions by email. Suggest one reason why this sampling method may lead to bias.

....

....

(1 mark)

2 Jamila is investigating the pay rises given to working adults in her town last year. The table below shows the number of working adults in Jamila's town.

Age (in years)	18-27	28-37	38-47	48-57	Over 57
No. of working adults	1200	2100	3500	3200	1500

Jamila plans to use stratified sampling to select a sample of 50 working adults from her town.

a) Suggest one reason why it might be sensible for Jamila to stratify her sample by age.

....

....

(1 mark)

b) Calculate how many people from each age group should be in the sample.

Remember to round decimals to the nearest whole number, and check your total is 50.

18-27 = ..., 28-37 = ..., 38-47 = ..., 48-57 = ..., Over 57 = ...

(3 marks)

c) Jamila wants to investigate the pay rises given to working adults across the UK last year. Can she use her sample data to draw conclusions about the whole population? Explain your answer.

....

....

(1 mark)

Statistical Hypothesis Testing

3 A study in the UK carried out in 2003 found that the median height of women was 163 cm. Adil wants to carry out a hypothesis test to investigate whether the proportion of women in the USA who were taller than 163 cm was different to the proportion in the UK in 2003.

a) Adil uses the large data set to randomly sample 30 women from the USA in 2003. 11 of the women had a height greater than 163 cm.
Use this information to carry out Adil's test at the 10% level of significance.

Since the median height was 163 cm, half of women in the UK were taller than 163 cm.

....

....

(6 marks)

b) Adil is going to do a different hypothesis test to investigate whether women's weights in the USA were different to women's weights in the UK in 2003. For this test, he is going to take a sample of 20 women from the large data set, chosen using systematic sampling. Explain how Adil could use systematic sampling to choose a sample from the 101 women in the large data set whose weight is recorded.

....

....

(3 marks)

4 Nate teaches judo classes at 'basic' and 'advanced' levels. Last year, 20% of his students had done judo for at least two years. Nate moves to a different judo club and claims that at this club, the percentage of his students who have done judo for at least two years is different.
To test this, he surveys a random sample of 20 of his new students.

a) Given that 7 of the sampled students have done judo for at least two years, use a binomial distribution to carry out a test of Nate's claim at the 5% significance level.

....

....

(6 marks)

b) State two assumptions that are needed for the binomial model you used in part a) to be valid.
Comment on whether each assumption is likely to be true for Nate's test.

....

....

....

....

(4 marks)

Statistical Hypothesis Testing

5 2016 records suggest that 45% of the members of a gym use the swimming pool. The gym's manager claims that the popularity of the swimming pool has decreased since then. He surveys a random sample of 50 members to test his claim.

After carrying out his test at the 5% level of significance, the manager concludes that there is evidence to suggest that the popularity of the pool has decreased.

Find the maximum possible number of gym members in the sample of 50 who use the pool.

....

(4 marks)

6 A bakery makes different flavours of muffins. The masses of their blueberry muffins can be modelled by a normal distribution with mean 110 g and standard deviation 3 g. A customer claims that the bakery's chocolate muffins weigh less than the blueberry muffins. A random sample of 15 chocolate muffins was found to have a mean mass of 108.5 g.

Assuming that the masses of the chocolate muffins can be modelled by a normal distribution with standard deviation 3 g, test the customer's claims at the 10% level of significance.

....

....

(6 marks)

7 The mean height of the sunflowers in a particular field is 150 cm. The heights of the sunflowers in a second field are known to follow a normal distribution, with a standard deviation of $\sqrt{20}$ cm. The mean height of a random sample of 6 of these sunflowers is 140 cm.

Test at the 1% level of significance whether the average height of the sunflowers in the second field is the same as for the sunflowers in the first field.

....

....

(6 marks)

Statistical Hypothesis Testing

8 The duration in minutes, X, of a car wash is normally distributed with mean 8 and standard deviation $\sqrt{1.2}$. After some maintenance work is carried out, the manager claims that the mean duration of the car wash has decreased. A hypothesis test is to be carried out to investigate this claim.

a) A random sample of 20 washes are timed and the mean duration is found to be 7.8 minutes. Carry out the test at the 5% significance level. You may assume that the standard deviation of the washes' durations is unchanged.

....

....

(6 marks)

b) Find the least value for the sample mean car wash duration that would have provided insufficient evidence to reject the null hypothesis for the test you carried out in part a). Give your answer to 3 significant figures.

....

(2 marks)

9 The average time taken by an employee at Sahil's company to get to work has previously been calculated to be 27 minutes. The journey times are assumed to be normally distributed. Sahil claims that the average journey time is now less than 27 minutes. He uses the journey times of 45 randomly-selected employees to calculate a sample mean of 24.8 minutes and a sample standard deviation of 6.1 minutes.

Use Sahil's sample data to test his claim at the 1% significance level.

Here you can use the sample variance as an estimate of the population variance.

....

....

(6 marks)

EXAM TIP

With hypothesis tests, think carefully about whether it's easier to find the p-value or the critical region — and make sure you're confident with your binomial and normal distribution calculator functions. And remember — always write a proper conclusion. For example, don't just say 'reject H_0', you also need to explain what rejecting H_0 means in the context of the question.

Score

58

Correlation and Regression

Correlation and regression are some of my favourite things. In fact, I even named my two dogs 'Correlation' and 'Regression' because I love these topics so much. My children Cora and Reggie thoroughly approved.

1 A construction company measures the length, y metres, of a cable when put under different amounts of tension, T kN (kilonewtons). The results of its tests are shown below.

T (kN)	1	2	3	5	8	10	12	15	20
y (metres)	3.05	3.1	3.13	3.15	3.27	3.4	3.1	3.5	3.6

a) Draw a scatter diagram to show these results.

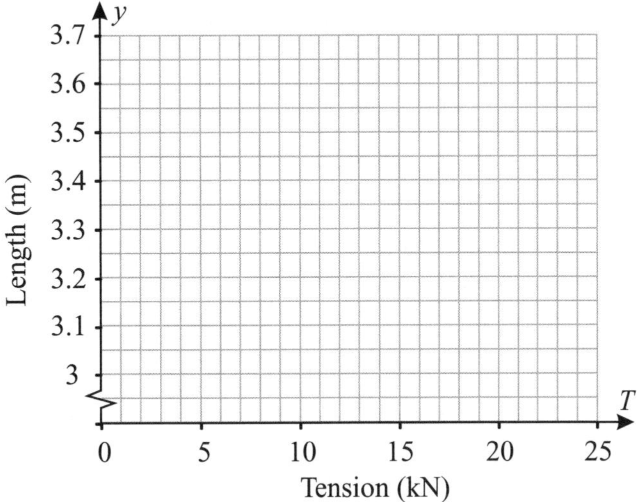

(2 marks)

b) One of the readings has been recorded inaccurately. Put a ring around this reading on your graph.

(1 mark)

c) Describe the correlation shown on your scatter diagram.

....

(1 mark)

An engineer believes a linear regression line of the form $y = a + bT$ could be used to accurately describe the results. She ignores the outlier, and calculates the equation of the regression line to be $y = 3 + 0.03T$.

d) Explain what these values of a and b represent in this context.

....

....

(2 marks)

e) Use the regression line to predict the length of the cable when put under a tension of 30 kilonewtons.

.... m

(1 mark)

f) Comment on the reliability of your estimate for part e).

....

....

(1 mark)

Correlation and Regression

2 Some ecologists carry out an investigation into the reindeer populations at different locations. They also record the human population density at the same locations. Their results are displayed in the scatter diagram below.

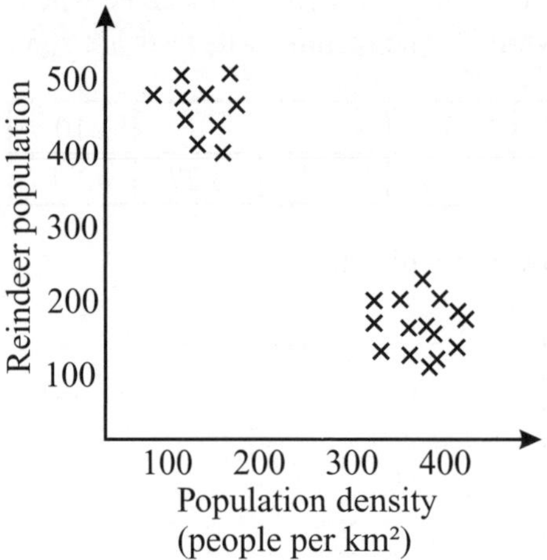

a) Jiao claims that there is negative correlation between human population density and reindeer population. Use the scatter diagram to comment on Jiao's claim.

....

....

....

(2 marks)

b) Killian claims that more people living in an area cause there to be fewer reindeer in that area. Do you agree with Killian's claim? Explain your answer.

....

....

....

(2 marks)

3 A student collects 10 pairs of data values from an estate agent's website. Each pair of values shows the floor area (A) and the advertised price (P) of a randomly-selected house.

Using her 10 pairs of values, the student calculates the correlation coefficient between A and P to be 0.634. The critical value for a 1-tail test for correlation at the 5% significance level for a sample size of 10 is 0.549. Explain whether or not there is evidence of positive correlation between A and P for the population of houses advertised on the whole website.

....

....

(2 marks)

Correlation and Regression

4 The scatter diagram below shows 'birth rate per 1000' and 'death rate per 1000' for European countries in the large data set.

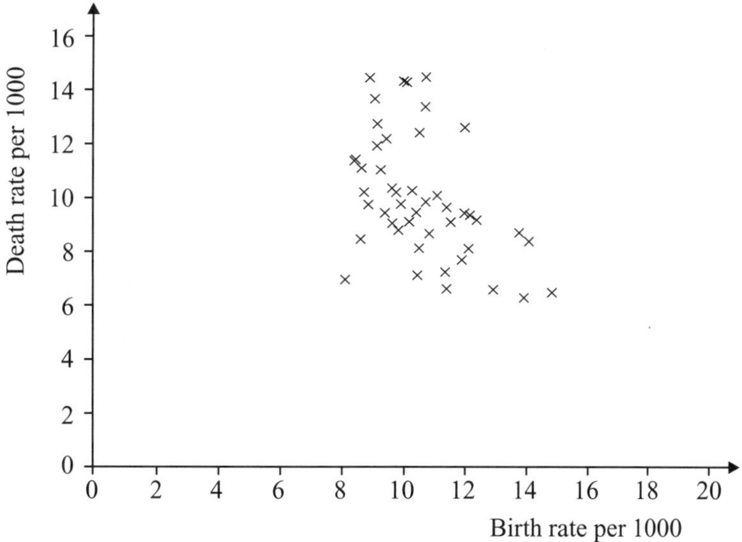

The data point for Monaco has been missed off the diagram. The birth rate per 1000 for Monaco is 6.65.

a) Give one reason why it might be difficult to accurately predict the death rate per 1000 for Monaco using the given scatter diagram.

....

....

(1 mark)

b) On the scatter diagram above, circle all the points corresponding to countries with a higher death rate per 1000 than birth rate per 1000.

(2 marks)

5 The ages, a, of 20 teenagers who belong to a swimming club and their personal-best times for swimming 100 metres, t, are recorded. A rank correlation coefficient between a and t is calculated to be –0.78.

Carry out a hypothesis test at the 1% significance level to investigate whether there is evidence of negative association between age and personal-best time. The critical value for the required test is –0.522.

....

....

(2 marks)

Scatter diagrams, regression lines and correlation coefficients help you to understand relationships between different sets of data — but just because two data sets are correlated, that doesn't necessarily mean that one causes changes in the other. Apparently, ice cream sales and murder rates are correlated... but that doesn't mean the next person you see eating ice cream is a killer.

Score

19

Kinematics 1

Kinematics is all about the way things move — by the end of this topic your flamboyant dance skills should be most impressive. For questions in this topic, give your answers to an appropriate degree of accuracy.

1 A particle moves in a straight line with a constant acceleration. The initial velocity of the particle is U ms^{-1} and its velocity at time T is V ms^{-1}, as shown on the velocity-time graph below.

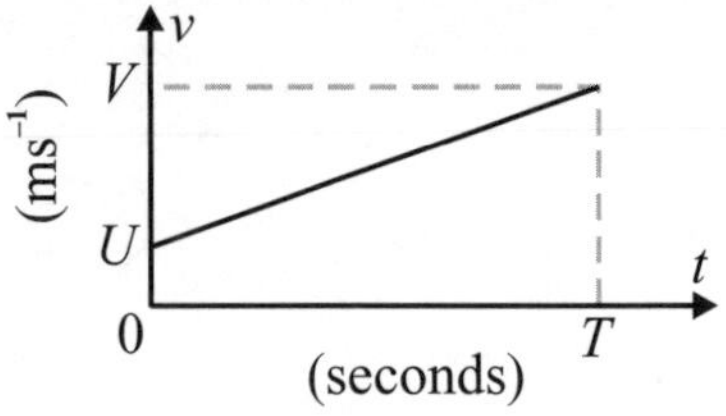

Using the graph, show that the displacement of the particle can be given by $S = \frac{1}{2}(U + V)T$.

....

(3 marks)

2 A ball is thrown vertically upwards with velocity 5 ms^{-1} from a point 2 m above the ground. The velocity-time graph below models the motion of the ball. Take g = 9.8 ms^{-2}.

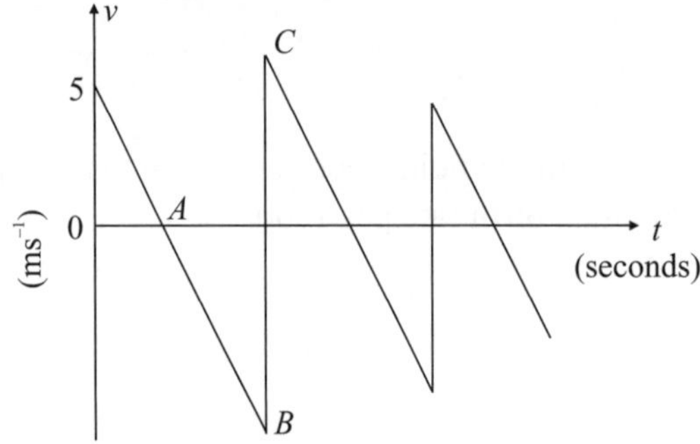

a) Find the time taken by the ball to reach point A.

....

(3 marks)

b) Find the velocity of the ball when it reaches point B.

....

(3 marks)

c) Explain why, in reality, this graph may not be an accurate model of the velocity of the ball.

....

(1 mark)

Kinematics 1

3 A train is moving along a straight horizontal track with constant acceleration. The train enters a tunnel which is 760 m long and exits 24 seconds later. 8 seconds before entering the tunnel, the train passes a sign. The sign is 110 m from the tunnel's entrance.

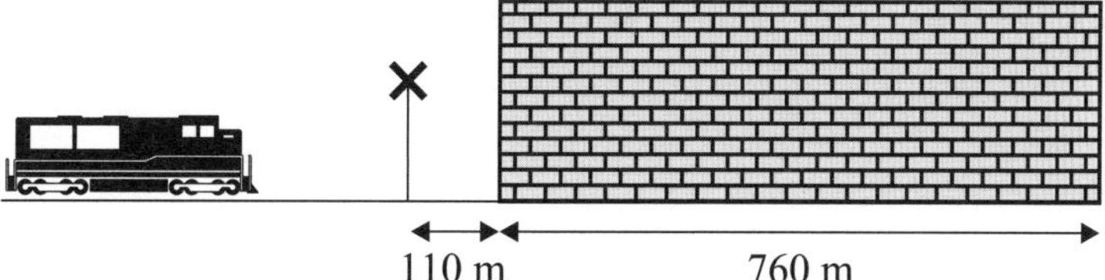

Find the train's acceleration and its speed when it passes the sign.

$a =$ .... $u =$ ....

(7 marks)

4 A robot initially has position vector $(2\mathbf{i} + \mathbf{j})$ m and is travelling with constant velocity $(\mathbf{i} + 3\mathbf{j})$ ms^{-1}. After 8 seconds it reaches point A. A second robot has constant velocity $(-5\mathbf{i} + 2\mathbf{j})$ ms^{-1} and takes 5 seconds to travel from point A to point B.

a) Calculate the difference in speed between the two robots.

....

(2 marks)

b) Find the direction of motion of the second robot, measured anticlockwise from the positive **i** direction.

....

(2 marks)

c) Find the position vectors of points A and B.

$A =$ ....

$B =$ ....

(2 marks)

Kinematics 1

5 At time $t = 0$, a particle, P, sets off from rest at the point O and accelerates uniformly. At time $t = 3$ seconds, the particle has position vector $(9\mathbf{i} - 18\mathbf{j})$ m relative to O.

a) Given that the particle continues to move with this constant acceleration, find an expression for the position vector, $\mathbf{p}$, of the particle at time t.

$\mathbf{p}$ = ...

(4 marks)

Also at time $t = 0$, a second particle, Q, moving with constant velocity $(3\mathbf{i} - 5\mathbf{j})$ ms^{-1} has position vector $(a\mathbf{i} + b\mathbf{j})$ m, where a and b are constants.

b) Given that the two particles collide at time $t = 8$ seconds, find the values of a and b.

a = ... b = ...

(3 marks)

6 A ball's velocity is modelled by the vector $\mathbf{v} = [(2t - 5)\mathbf{i} + (3t - 8)\mathbf{j}]$ ms^{-1}, where $\mathbf{i}$ and $\mathbf{j}$ are unit vectors pointing east and north respectively. At time $t = 2$ s, the ball is at position vector $(-2\mathbf{i} + 10\mathbf{j})$ m.

a) Find the value of t at which the ball is travelling directly eastwards.

t = ...

(2 marks)

b) Calculate the range of t for which the ball is to the west of the origin.

....

(5 marks)

Kinematics 1

7 A particle P sets off from the origin at $t = 0$ and starts to move along the x-axis in the direction of increasing x. After t seconds, P has velocity v ms^{-1}, where v is given by:

$$v = \begin{cases} 11t - 2t^2 & 0 \le t \le 5 \\ 25 - 4t & t > 5 \end{cases}$$

a) Find the displacement of P from the origin at $t = 5$.

....

(4 marks)

b) Find the time taken for P to return to the origin.

....

(5 marks)

8 An object has constant acceleration a and an initial velocity of u.

Given that the object sets off from the origin, use integration to show that the object's displacement at time t is given by the formula $s = ut + \frac{1}{2}at^2$.

(4 marks)

Kinematics 1

9 A particle moves in a straight line, beginning at the origin at $t = 0$.
Its velocity, v ms^{-1}, at time t seconds is given by $v = 2t - 3e^{-2t} + 4$.

a) Find an expression for the particle's acceleration, a ms^{-2}, at time t.

$a =$ ....

(2 marks)

b) Find the range of values for the particle's acceleration.

....

(3 marks)

c) Find an expression for the particle's displacement from O, s m, at time t.

$s =$ ....

(3 marks)

10 A particle is moving in a curved path. Its velocity, $\mathbf{v}$ ms^{-1}, at time t seconds is given by
$\mathbf{v} = (2\cos 3t + 5t)\mathbf{i} + (2t - 7)\mathbf{j}$, where the unit vectors $\mathbf{i}$ and $\mathbf{j}$ are in the directions of east and north respectively.

a) Find an expression for the particle's acceleration, $\mathbf{a}$ ms^{-2}, after t seconds.

$\mathbf{a} =$ ....

(2 marks)

b) Calculate the magnitude of the particle's maximum acceleration. Give an exact value.

....

(3 marks)

Kinematics questions can be tough, because they combine lots of skills — you need to be comfortable with using vectors, differentiating and integrating, as well as all the mechanics-y stuff. Just remember, whenever you see 'constant' or 'uniform' acceleration in an exam question you should think 'suvat'. If acceleration isn't constant, then it's time to whip out the calculus.

Score

63

Kinematics 2

These questions are all about projectiles. Master the maths of trajectories and you'll be a pro-golfer in no time. Give your answers to an appropriate degree of accuracy and take $g = 9.8$ ms^{-2}, unless told otherwise. Fore!

1 A golf ball is struck from a point T on horizontal ground with velocity $(29\mathbf{i} + p\mathbf{j})$ ms^{-1}, at an angle α to the horizontal. The ball takes 5 seconds to land on the ground at point G.

a) Find p.

p = ...

(2 marks)

b) Find the horizontal distance TG.

(2 marks)

2 A stone is thrown upwards from point A on a cliff, 22 m above horizontal ground, with speed 14 ms^{-1} at an angle of 46° to the horizontal. After projection, the stone moves freely under gravity and lands at B on the horizontal ground, as shown below.

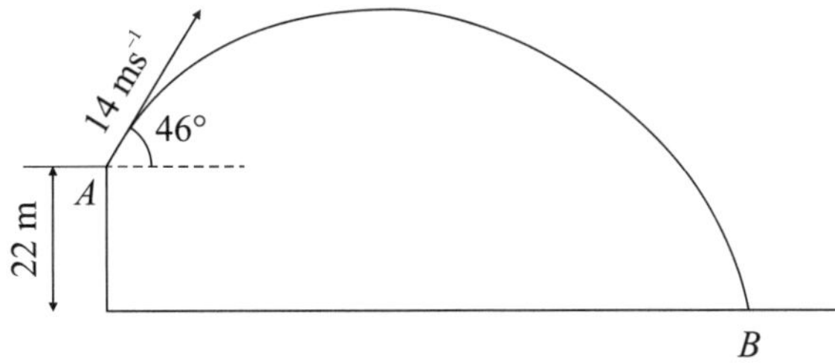

a) Find the length of time for which the stone is at least 22 m above the ground.

....

(4 marks)

b) Calculate the speed of the stone as it hits the ground.

....

(5 marks)

Kinematics 2

3 A stone is thrown from A, at a height of 1 m above the ground. The stone's initial velocity is 10 ms^{-1} at an angle of 20° above the horizontal. The stone lands on the horizontal ground at B.

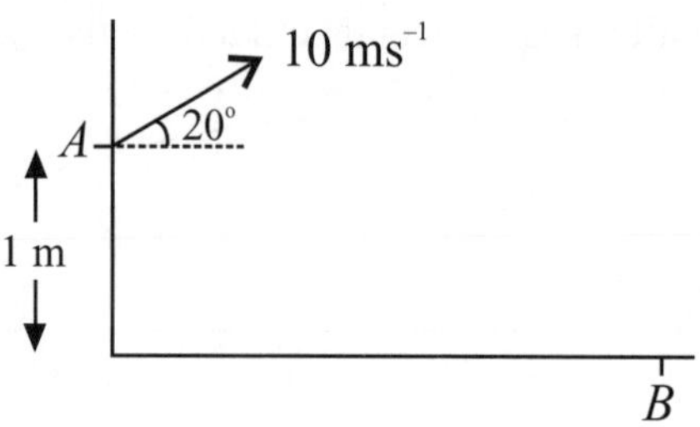

a) Calculate the maximum height above the horizontal ground that the stone reaches during its flight.

....

(4 marks)

b) Find the time it takes the stone to travel from A to B.

....

(3 marks)

c) Explain why the answer you found in part b) might not be the actual time it takes the stone to travel from A to B.

....

....

(1 mark)

Projectiles questions can look really complicated. The trick is: take a deep breath, write out the variables you've been given in the question and then think which suvat equation will let you work out the next bit of the question. Sometimes, you'll have to go through this process a couple of times before you get to the final answer, but so long as you're careful you'll get there.

Score

21

Forces and Newton's Laws

Try as you might, there's no escaping Newton's laws — they pretty much govern the way everything moves. In this topic, you should give your answers to an appropriate degree of accuracy and take $g = 9.8 \text{ ms}^{-2}$.

1 Three forces act at a point, O, as shown below.

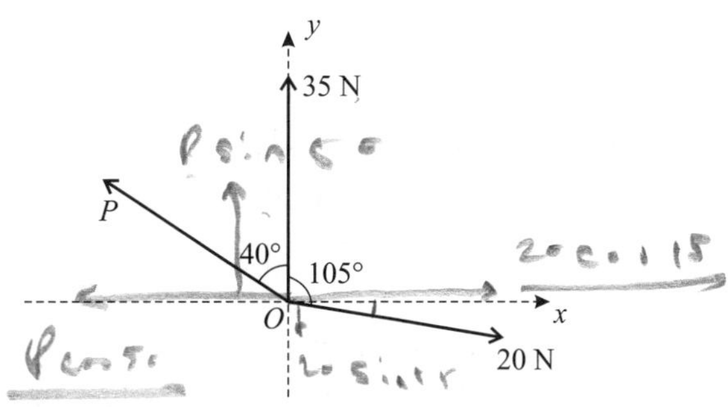

a) Given that there is no resultant force in the direction of the x-axis, find P.

P = ...

(2 marks)

b) Find the magnitude and direction of the overall resultant force, R.

....

(3 marks)

2 Two forces, $\mathbf{A} = (2\mathbf{i} - 11\mathbf{j})$ N and $\mathbf{B} = (7\mathbf{i} + 5\mathbf{j})$ N, act on a toy helicopter of mass 500 g.

a) Find the magnitude of the resultant force.

....

(3 marks)

b) Find the acceleration of the helicopter, $\mathbf{a}$ ms^{-2}.

$\mathbf{a}$ = ...

(2 marks)

Forces and Newton's Laws

3 A 2 kg ring is threaded on a rough horizontal rod. A rope is attached to the ring, and is held at 40° below the horizontal, as shown. The normal contact force of the rod on the ring is R N, the tension in the rope is S N and frictional force between the rod and the ring is F N.

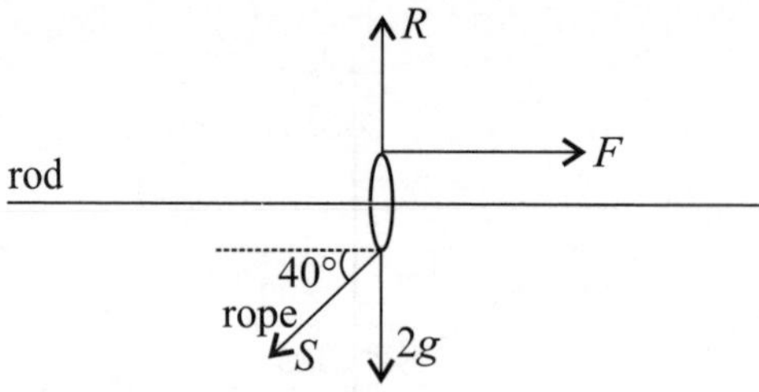

The coefficient of friction between the rod and the ring is 0.3.
Given that the ring is stationary, find the range of possible values that S can take.

S = ...

(5 marks)

4 A horizontal force of 25 N causes a particle of mass 7 kg to accelerate up a rough plane inclined at 15° to the horizontal, with acceleration of magnitude 0.2 ms^{-2}, as shown on the diagram to the right.

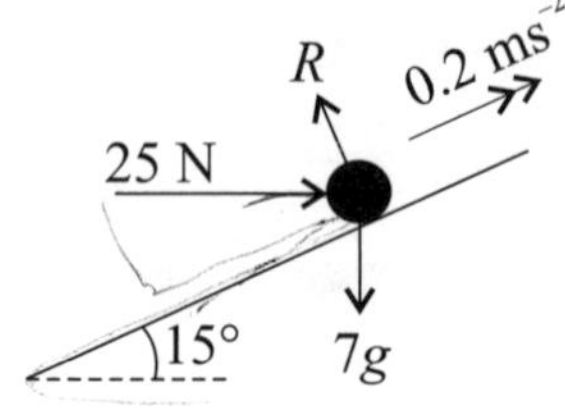

a) Calculate the coefficient of friction between the mass and the plane to 2 decimal places.

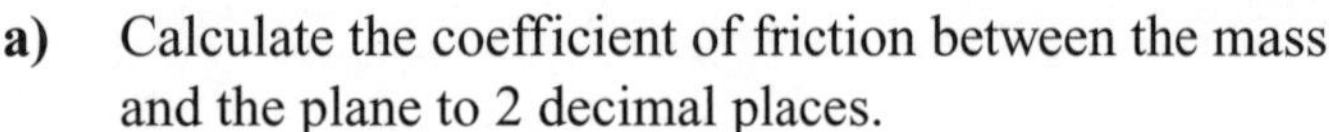

μ = ...

(6 marks)

b) The 25 N force is now removed and the particle is released from rest.
Find how long the particle takes to slide a distance of 3 m down the plane.

....

(6 marks)

Forces and Newton's Laws

5 A block of mass m kg rests on rough horizontal ground. The block is held in equilibrium by a string under tension T N pulling upwards at an angle of 51.3° to the horizontal. The coefficient of friction between the block and the ground is 0.6 and the magnitude of the frictional force is 1.5 N.

a) Sketch a diagram showing the block and the forces acting upon it.

(1 mark)

b) Given that friction is limiting, find the tension in the string, T, and the mass of the block, m.

$T =$ ....

$m =$ ....

(5 marks)

6 A ball of weight W N is attached to a string under tension T N on a rough plane inclined at β to the horizontal. The string is parallel to the plane. The ball is held in equilibrium by a frictional contact force F N, acting in the same direction as T.

Assuming that T acts parallel to the plane, and friction is limiting,
show that the coefficient of friction $\mu = \tan\beta - \frac{T}{W}\sec\beta$.

(5 marks)

Forces and Newton's Laws

7 A woman is travelling in a lift. The lift is rising vertically and is accelerating at a rate of 0.75 ms^{-2}. The lift is pulled upwards by a light, inextensible cable. The tension in the cable is T N and the lift has mass 500 kg. The floor of the lift exerts a force of 675 N on the woman. Find T.

$T =$ ....

(4 marks)

8 One end of a light inextensible string is attached to a block A, of mass 35 kg. The string passes over a fixed, smooth pulley and is attached at the other end to a block B, of mass M kg.

The system is held at rest by a vertical light inextensible string attached to B at one end and to horizontal ground at the other, as shown. The tension in this second string is K N.

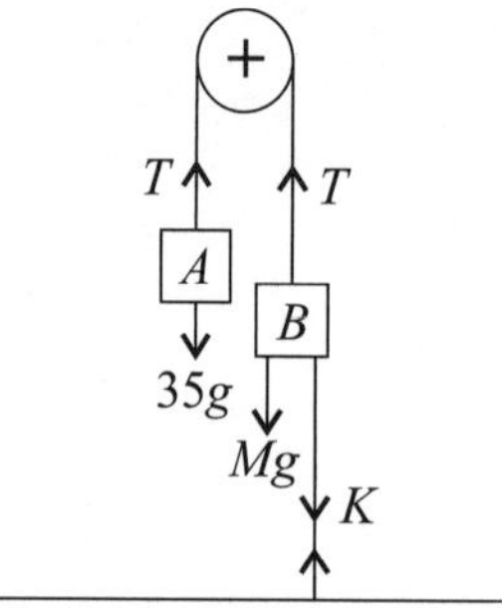

a) Find an expression for K in terms of M and g.

....

(2 marks)

b) **(i)** The string fixing B to the ground is cut. Three seconds after the string is cut, A is travelling at 1 ms^{-1} towards the ground. Calculate M.

$M =$ ....

(5 marks)

(ii) Give one assumption you have made about the behaviour of the system after the string is cut.

....

(1 mark)

Forces and Newton's Laws

9 A particle, A, is attached to a weight, W, by a light inextensible string which passes over a smooth pulley, P, as shown. When the system is released from rest, with the string taut, A and W experience an acceleration of 4 ms^{-2}. A moves across a rough horizontal plane and W falls vertically. The mass of W is 1.5 times the mass of A.

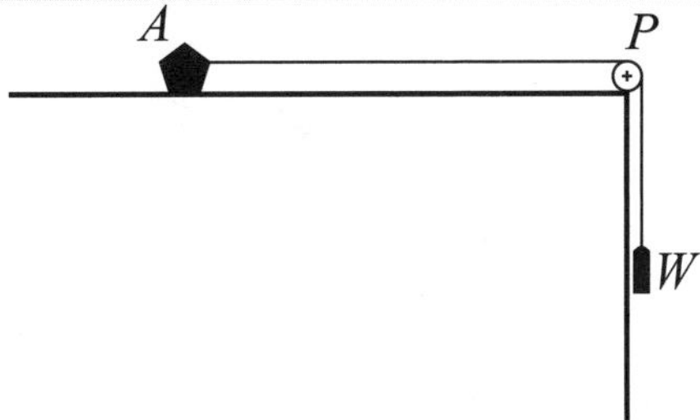

a) Given that the mass of A is 0.2 kg, find the coefficient of friction, μ, between A and the horizontal plane.

μ = ...

(6 marks)

W falls for h m until it impacts the ground and does not rebound. A continues to move until it reaches P with speed 3 ms^{-1}. The initial distance between A and P is $\frac{7}{4}h$.

b) Find the time taken for W to impact the ground.

....

(7 marks)

c) How did you use the information that the string is inextensible?

....

....

(1 mark)

Forces and Newton's Laws

10 A particle A of mass 3 kg is placed on a rough plane inclined at an angle of $\theta = \tan^{-1}\frac{3}{4}$ above the horizontal and is attached by a light inextensible string to a second particle B of mass 4 kg.

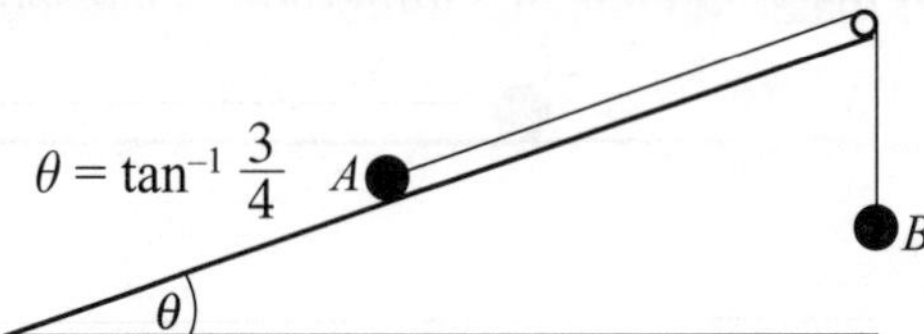

The string passes over a smooth pulley at the top of the inclined plane so that particle B hangs freely. When the system is released, particle B moves downwards vertically with an acceleration of 1.4 ms^{-2} and A moves up the plane.

a) Find the coefficient of friction between particle A and the plane.

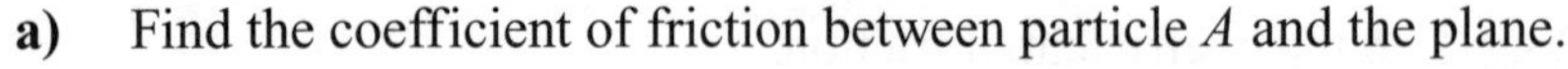

$\mu =$ ....

(7 marks)

Two seconds after the particles are released, the string breaks. A then continues up the plane until it comes instantaneously to rest. At no point does A reach the pulley.

b) Find how far particle B moves from the start of the motion until the string breaks.

....

(2 marks)

c) Calculate how far particle A moves from the instant the string breaks until it comes to rest.

....

(5 marks)

Never in the history of mathematics has drawing a detailed diagram been more important. You don't stand a chance of keeping all the numbers involved in these questions in your head. For connected-particle questions, you might even want to draw more than one diagram — one for each particle. Trust me on this one, a sketch is worth a thousand words.

Score

78

Moments

It had to happen sooner or later, and there's no putting it off any longer — it's your favourite topic... moments. As ever, give your answers to an appropriate degree of accuracy and take $g = 9.8 \text{ ms}^{-2}$, unless told otherwise.

1 A 6 m long uniform beam of mass 20 kg is in equilibrium. One end is resting on a vertical pole, and the other end is held up by a vertical wire attached to that end so that the beam rests horizontally. There are two 10 kg weights attached to the beam, situated 2 m from either end.

Find, in terms of g:

a) T, the tension in the wire.

....

(2 marks)

b) R, the normal reaction at the pole.

....

(2 marks)

2 A uniform lamina, $ABCD$, of weight 8 N and dimensions 3 m × 1 m, is attached to a wall at its corner A with a hinge. The lamina is held in equilibrium by a vertical force F acting at point C as shown.

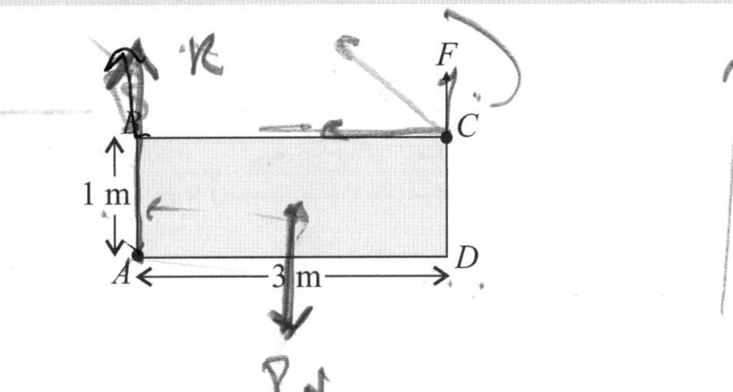

Find the magnitude of F.

....

(3 marks)

3 A uniform rod AB of length 4 m and weight 20 N hangs in equilibrium in a horizontal position supported by two vertical inextensible strings attached at A and B. A bead of weight 16 N rests on the rod at a point C, which is 3 m from A. The bead can be modelled as a particle.

a) Find the tensions in the strings at A (T_1) and at B (T_2).

T_1 = 14N

T_2 = 22N

(3 marks)

b) The bead is now moved along the rod so that the tension in the string at B is twice the tension in the string at A. Find the distance of the bead from A.

....

(3 marks)

Moments

4 A non-uniform wooden beam, LM, has weight W N and length 2 m. A painter places the beam on supports at P and Q, as shown below. The painter has weight $5W$ and wants to stand on the beam.

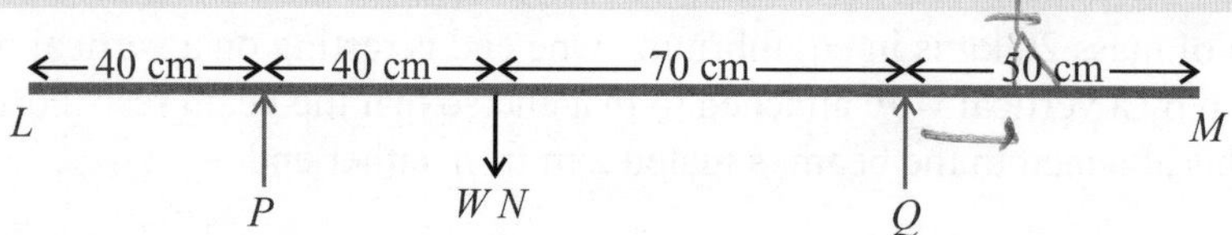

Find the range of points on the beam that could support the painters full weight without the beam tipping.

....

(6 marks)

5 A skateboarder balances a skateboard on a rail. The skateboard has mass 4 kg and can be modelled as a uniform horizontal rod, AB, 0.8 m in length. The skateboard is in contact with the rail at point C, where $AC = 0.5$ m. The skateboarder has mass 80 kg and places her feet at points X and Y. The skateboard is in equilibrium when the force applied at Y is three times the force applied at X and CY is half the magnitude of CB.

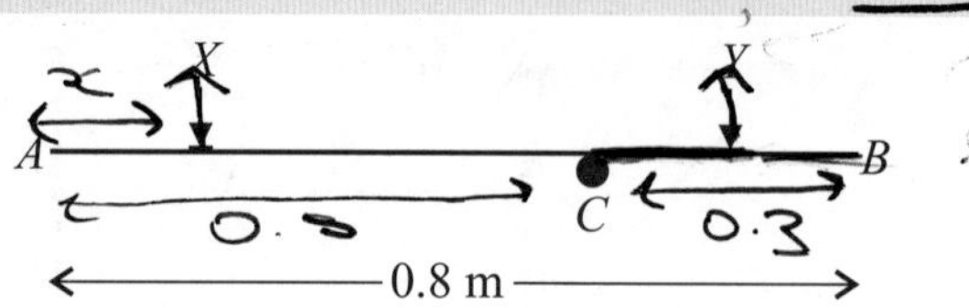

a) Find the distance AX.

....

(5 marks)

b) Find the magnitude of the normal reaction, R, at the rail.

....

(2 marks)

c) An object of mass 3 kg is placed on the skateboard at point A. The skateboarder moves her foot such that AX is increased by 13 cm. The skateboard remains horizontal and in equilibrium. Find the magnitude of the forces applied at X (F_X) and Y (F_Y).

F_X = ...

F_Y = ...

(5 marks)

If I had to sum up this topic in 5 words, I'd say this: "Equate moments and balance forces." Okay, fair enough, you have to be able to actually do the equating of moments and balancing of forces, and if I had about 800 more words to play with, I'd talk you through that. But used wisely I reckon those 5 words will get you around 80% of the marks on these pages.

Score

31

General Certificate of Education
Advanced Level

A-Level Mathematics
Practice Exam Paper 1: Pure Mathematics and Mechanics

Time Allowed: 2 hours

There are 100 marks available for this paper.

Formulas are given on pages 164 to 166.

Section A (25 marks)

1 $f(x) = x^3 - 2x^2 - 13x - 10$

a) Show that $(x + 1)$ is a factor of $f(x)$.

(2 marks)

b) Hence, or otherwise, solve $f(x) = 0$.

(3 marks)

2 **In this question you must show detailed reasoning.**

Find the coordinates of the points of intersection of the circle $(x + 1)^2 + (y - 2)^2 = 5$ and the line $x - 2y + 5 = 0$.

(5 marks)

3 For small values of θ, show that $\frac{1 - \tan 3\theta}{\cos 4\theta - \sin \theta^2} \approx \frac{1}{1 + 3\theta}$.

(3 marks)

4 $f(x) = \frac{2}{x^2}$

a) Sketch the graph of $y = f(x)$, giving the equations of any asymptotes.

(3 marks)

b) Sketch the graph of $y = -f(x) + 1$, giving the equations of any asymptotes.

(3 marks)

5 Find $\int \frac{15x^2 + 4x}{5x^3 + 2x^2 + 6}\,dx$.

(3 marks)

6 For this question, take $g = 9.8 \text{ ms}^{-2}$.

A uniform rod, AB, of length 6 m and mass 6 kg rests on two supports, P and Q, as shown.

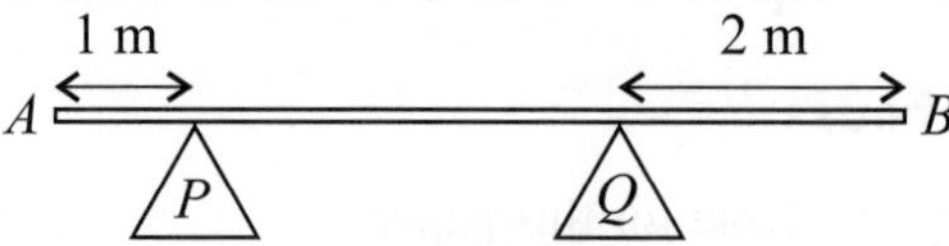

A 4 kg particle is placed on the rod at a distance of x m from A.
Given that the reaction forces on the rod at P and Q are now equal, and that the rod is in equilibrium, find the value of x.

(3 marks)

Section B (75 marks)

7 A particle P of mass 2.5 kg is moving in a horizontal plane under the action of a single force, **F** newtons. At t seconds, the position vector of P is **r** m, where **r** is given by:

$$\mathbf{r} = (t^3 - 6t^2 + 4t)\mathbf{i} + (7t - 4t^2 + 3)\mathbf{j},$$

where **i** and **j** are the unit vectors directed due east and due north respectively.

a) Show that when $t = 5$, the velocity of P is $19\mathbf{i} - 33\mathbf{j}$.

(3 marks)

b) Find the value of t when P is moving due south.

(3 marks)

c) Find the magnitude of the resultant force acting on P when $t = 3$.

(5 marks)

8 The functions f, g and h are defined below.

$\text{f}(x) = \dfrac{2x+7}{3x-5}, x \in \mathbb{R}, x \neq \dfrac{5}{3}$ $\quad$ $\text{g}(x) = x^2 - k, x \in \mathbb{R}$ where k is a positive constant $\quad$ $\text{h}(x) = \cos x, x \in \mathbb{R}$

a) Find $\text{f}^{-1}(x)$.

(2 marks)

b) $\text{gf}(2) = 120$. Find the value of k.

(3 marks)

c) Find $\text{gh}(x)$ and $\text{hg}(x)$, and hence show that the range of $\text{gh}(x)$ is not the same as the range of $\text{hg}(x)$.

(4 marks)

9 **a)** Show that the x-coordinate of the point of intersection of the curves $y = \frac{1}{x^3}$ and $y = x + 3$ satisfies the equation $x^4 + 3x^3 - 1 = 0$.

(1 mark)

$x^4 + 3x^3 - 1 = 0$ has a root α in the interval [0.5, 1]

b) Apply the Newton-Raphson method to obtain a second and third approximation for α, using 0.6 as the first approximation for α. Give your answers to 4 decimal places.

(3 marks)

10 A helicopter E is modelled as a particle. It sets off from its base O on the coast and flies at constant height with initial velocity $(40\mathbf{i} + 108\mathbf{j})$ kmh^{-1} and constant acceleration $(4\mathbf{i} + 12\mathbf{j})$ kmh^{-2}, where **i** and **j** are the unit vectors in the direction of east and north respectively. Thirty minutes after E has left the base, the pilot informs the base that he has engine failure and is ditching into the sea.

a) Find the position vector of E relative to O at the time it experiences engine failure.

(3 marks)

The base immediately informs another helicopter, F, that E is in need of assistance. F is modelled as a particle flying towards E's position at a constant height from position $(2.5\mathbf{i} + 25.5\mathbf{j})$ km, with constant acceleration and initial velocity $(60\mathbf{i} + 120\mathbf{j})$ kmh^{-1}.

b) Given that F takes 15 minutes to reach E, find the magnitude of the acceleration of F.

(3 marks)

The base sends out a third helicopter, G, on another mission. G is modelled as a particle, and leaves the base O with velocity $(30\mathbf{i} - 40\mathbf{j})$ kmh^{-1} and accelerates at a constant $(5\mathbf{i} + 8\mathbf{j})$ kmh^{-2} towards its destination.

c) Find the position vector of G when it is travelling due east.

(6 marks)

11 The curve C has parametric equations:

$$x = t^3 + 2, \ y = t^2 + 2$$

The point P lies on the curve where $t = 1$. The line l is the tangent to C at P.

Find the Cartesian equation of the line l in the form $y = mx + c$.

(5 marks)

12 **a)** Find the coordinates of the stationary point on the curve $y = x \ln x$.
Give your answer in exact form.

(5 marks)

b) Determine the nature of the stationary point. Give a reason for your answer.

(2 marks)

13 The rate of growth of a population of bacteria is proportional to the number of bacteria in the population.

a) Show that the number of bacteria in the population, P, can be modelled by the function $P = Qe^{kt}$, where t is the time in hours after the population was first observed and Q and k are constants.

(3 marks)

b) When the experiment begins, there are 5300 bacteria. After 6 hours, there are 876 bacteria present.
Find k to 1 decimal place.

(3 marks)

c) Use your value of k to describe how the population of bacteria is changing.

(1 mark)

14 A golf ball is hit off a rock of height 0.5 m with initial velocity U ms^{-1} at an angle α° above the horizontal.
The golf ball lands on horizontal ground at point X, as shown.

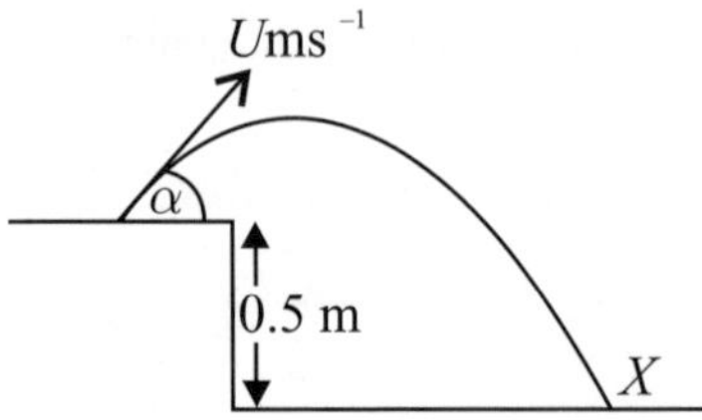

a) Show that the golf ball reaches maximum height $h = \dfrac{g + U^2 \sin^2 \alpha}{2g}$ m above the ground, where g is acceleration due to gravity.

(4 marks)

b) Show that V, the speed of the golf ball when it lands at X, is $V = \sqrt{U^2 + g}$ ms^{-1}.

(5 marks)

A beach ball is projected from the same point as the golf ball, with velocity U ms^{-1} and angle of projection α.

c) Suggest one way the model used in parts a) and b) could be adapted for the beach ball.

(1 mark)

15 For this question, take $g = 9.8 \text{ ms}^{-2}$.

Two particles, A and B, are connected by a light inextensible string that passes over a smooth pulley. A rests on a smooth plane inclined at an angle of $\theta°$ to the horizontal, where $\theta = \tan^{-1}\left(\frac{3}{4}\right)$, and B rests on a rough plane inclined at 70° to the horizontal, as shown. The coefficient of friction between B and the plane is 0.45. A has mass m kg and B has mass $2m$ kg.

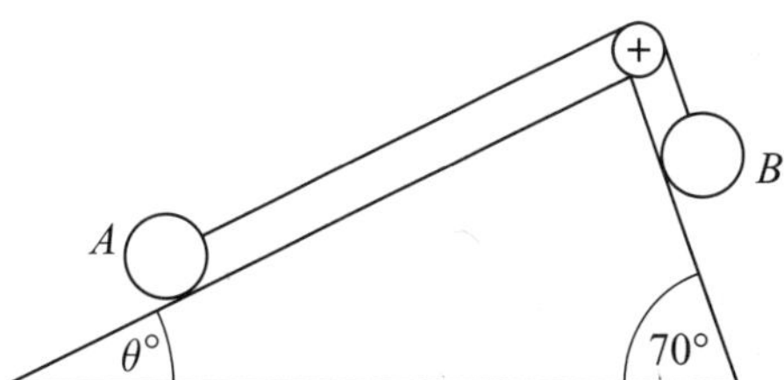

The system is released from rest and B starts to move down the plane.

a) Find the acceleration of particle A towards the pulley.

(6 marks)

b) The mass of A is increased to 10 kg. It is now on the point of sliding down the plane. Find the mass of B.

(4 marks)

END OF EXAM PAPER

TOTAL FOR PAPER: 100 MARKS

General Certificate of Education
Advanced Level

A-Level Mathematics
Practice Exam Paper 2: Pure Mathematics and Statistics

Time Allowed: 2 hours

There are 100 marks available for this paper.

Formulas are given on pages 164 to 166.

Section A (23 marks)

1 $f(x) = x^2 - 10x + 14$.
Write $f(x)$ in the form $(x + p)^2 + q$ and hence find the equation of the line of symmetry of the graph of $y = f(x)$.

(3 marks)

2 Write $2 \log a^3b - \log ab$ as a single logarithm in its simplest form.

(3 marks)

3 Chloe visits her local post office each weekday.
The probability that she has to queue for longer than 2 minutes on any visit is 0.4.
Find the probability that, in 50 visits, Chloe has to queue for longer than 2 minutes on exactly 20 occasions.

(2 marks)

4 Find the range of values of x for which $|4x + 3| < 11$.

(3 marks)

5 Prove that $\sec 2\theta \equiv \dfrac{\sec^2 \theta}{2 - \sec^2 \theta}$

(3 marks)

6 Find $f'(x)$ for $f(x) = \dfrac{\sin x}{(2x + 1)^4}$.

(4 marks)

7 **In this question you must show detailed reasoning.**

Evaluate the definite integral $\displaystyle\int_5^{10} \frac{2x}{\sqrt{x - 1}}\, dx$, using the substitution $u = \sqrt{x - 1}$.

(5 marks)

Section B (77 marks)

8 A researcher wants to collect data from two towns (A and B) in his local area.

In town A, 1524 houses are valued at below £150 000, 4279 houses are valued between £150 000 and £250 000 and 851 houses are valued at above £250 000.

a) Explain how the researcher could use stratified sampling to choose a sample of 100 households, stratified by house value, from town A.

(3 marks)

After collecting his data, the researcher wants to use it to investigate the weekly amount of money spent on food (x, in £) by households in these two towns. From the sample of households from town A, the researcher makes the following table:

Weekly amount of money spent on food (x, in £)	Frequency, f
$0 \le x < 30$	3
$30 \le x < 50$	12
$50 \le x < 75$	26
$75 \le x < 100$	31
$100 \le x < 140$	23
$140 \le x < 200$	5

b) Estimate the sample mean and sample standard deviation of the weekly amount spent on food in town A.

(3 marks)

For his sample from town B, the researcher calculates that, for the weekly amount spent on food, the sample mean is £61.08 and the sample standard deviation is £19.48.

c) Compare the weekly amount of money spent on food for towns A and B.

(2 marks)

9 Adam usually goes to the gym on Saturdays. When he does, he uses only the swimming pool or the rowing machines. On any given Saturday:

R represents the event that he uses the rowing machines.

S represents the event that he uses the swimming pool.

$$P(R \cup S) = \frac{8}{10} \quad P(R|S) = \frac{4}{9} \quad P(S|R) = \frac{4}{11}$$

a) Calculate P(R) and P(S).

(5 marks)

b) Explain whether the events R and S are independent.

(2 marks)

T represents the event that Adam has a takeaway on any given Saturday.

The events S and T are independent.

$$P(T) = 0.15 \quad P(R \cap T) = 0.08 \quad P(R \cap S \cap T) = 0.05$$

c) Draw a Venn diagram to show events R, S and T, giving the probabilities for each region.

(4 marks)

10 Alice and Ben are studying the approximate number of a particular bacteria, y, in ponds of different maximum depths, x m. They collect data from 8 ponds, with maximum depths varying from 0.5 m to 12 m, and calculate the correlation coefficient between x and y for their data. Alice obtains a value of $r_A = 0.71$, and Ben calculates his value to be $r_B = 1.09$.

a) Interpret, in context, the value of Alice's correlation coefficient, and explain how you can tell that Ben must have made a mistake in his calculation.

(2 marks)

b) Alice claims that her results show that deeper ponds result in higher numbers of bacteria. Comment on this claim.

(1 mark)

c) Stating your hypotheses clearly, test at the 5% significance level whether Alice's results are statistically significant evidence of positive correlation between the number of this type of bacteria and pond depth in the population of ponds. The critical value for this test is 0.6251.

(2 marks)

11 A juice manufacturer produces orange juice in cartons. The volume of juice in a carton, J ml, is assumed to be normally distributed. Based on previous data, the quality control officer finds that 5% of cartons contain less than 493 ml, and 2.5% contain more than 502 ml.

a) Find the mean and standard deviation of J.

(5 marks)

A carton meets the manufacturer's standards if it contains more than 492 ml of juice.

b) Find the probability of a carton failing to meet the manufacturer's standards.

(2 marks)

In an effort to cut costs, the manufacturer modifies the production process. After this process has been in place for a month, a random sample of 20 cartons are found to contain a mean volume of 498.7 ml of juice.

c) Test, at a 1% level of significance, whether or not the mean volume of juice in a carton has changed under the new production process. You must state your hypotheses clearly.

(6 marks)

12 In trapezium $ABCD$, the acute angle $BAD = x$ radians. Angles BCD and ABC are $\frac{\pi}{2}$. $DA = 5$ cm and $AB = 2CD$, as shown in the diagram.

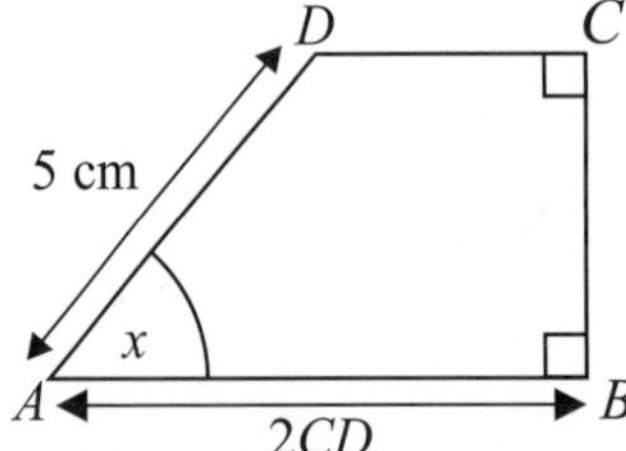

a) Show that an expression for the perimeter, P cm, can be written as $P = R\sin(x + \alpha) + 5$, where R and α are values to be found.

(7 marks)

b) Given that the perimeter of the trapezium is 17 cm, find the value of x to 3 significant figures.

(3 marks)

13 The graphs below show data about birth rates (per 1000 people) from the large data set.

Figure 13.1 shows the birth rates for 200 countries.

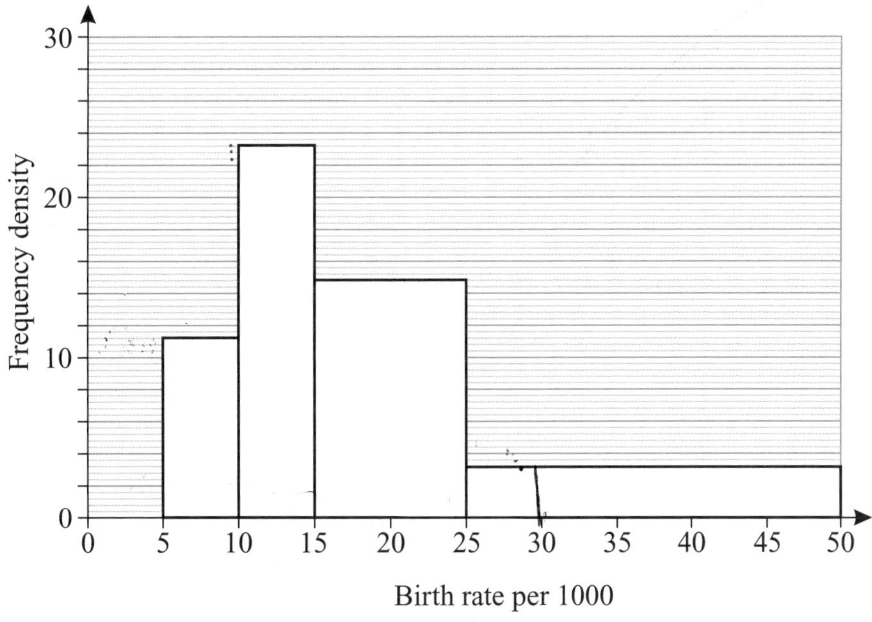

Figure 13.1

Figure 13.2 shows the birth rates for countries in Africa and Europe.

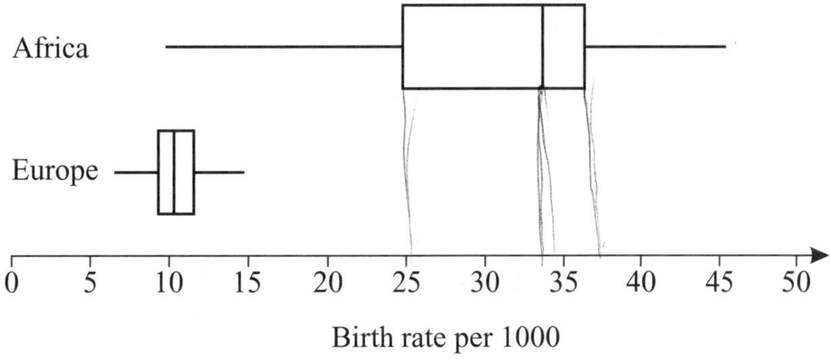

Figure 13.2

a) Eritrea has a birth rate (per 1000 people) of 30. Use Figure 13.1 to estimate the number of countries shown on the graph that have a higher birth rate than Eritrea.

(3 marks)

b) **(i)** Compare the distributions of birth rates for countries in Africa and Europe using Figure 13.2.

(3 marks)

(ii) Hashim suggests modelling the data for African countries in Figure 13.2 using a normal distribution. Explain why this would not be an appropriate model.

(1 mark)

Below is some information about the birth rates for countries in the Middle East:

minimum = 9.84, mean = 19.75, standard deviation = 6.22, maximum = 31.45.

c) Show that there are no outliers for the countries in the Middle East.

(3 marks)

Figure 13.3 shows the relationship between birth rates (B, per 1000) and GDP per capita (G, US \$, 1000s) for countries in Africa.

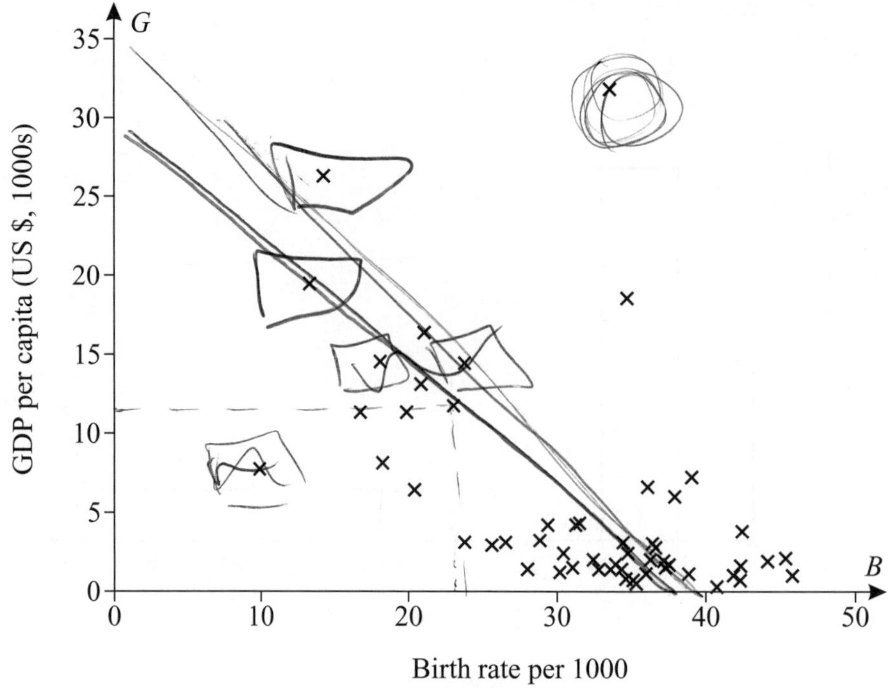

Figure 13.3

The regression line for the scatter graph is calculated to be:

$$G = 19.438 - 0.438B$$

d) **(i)** Describe the type of correlation shown in Figure 13.3.

(1 mark)

(ii) Figure 13.3 has the data value for Swaziland missing. It is known that the birth rate (per 1000) in Swaziland is 24.67. Use the regression line to estimate the GDP per capita for Swaziland.

(2 marks)

(iii) Comment on the reliability of the estimate for Swaziland in part (ii).

(1 mark)

Country A has a life expectancy close to the overall mean for all African countries, but is one of the largest oil producers in Africa. Countries B and C are islands that are popular luxury tourist destinations.

e) **(i)** Identify Country A by drawing a circle around it on Figure 13.3.

(1 mark)

(ii) Identify Countries B and C by drawing squares around them on Figure 13.3.

(1 mark)

14 A curve has equation $x^2 - xy = 2y^3$. Find the gradient of the curve at the point (2, 1).

(5 marks)

15 A pan containing soup at a temperature of 95°C is removed from a stove and the soup is left to cool in a kitchen. t minutes after being removed from the stove, the temperature of the soup is T °C.

The rate of heat loss of the soup over time is modelled as being directly proportional to the difference in temperature between the soup and the kitchen.
The temperature of the kitchen is 15°C.

a) Explain why $\frac{dT}{dt} = -k(T - 15)$, for some positive constant k.

(1 mark)

b) Find an expression for T in terms of t and k.

(3 marks)

After 10 minutes, the temperature of the soup is 55°C.

c) Find the exact value of k, and hence show that $T = 80(0.5)^{pt} + 15$ where p is a positive constant to be found.

(4 marks)

d) State one assumption that has been made for this model.

(1 mark)

END OF EXAM PAPER

TOTAL FOR PAPER: 100 MARKS

General Certificate of Education
Advanced Level

A-Level Mathematics
Practice Exam Paper 3: Pure Mathematics and Comprehension

Time Allowed: 2 hours

There are 75 marks available for this paper.

Formulas are given on pages 164 to 166.

Section A (60 marks)

1 The diagram shows a section of the graph of $y = 2\cos\frac{x}{4}$.
It crosses the x-axis at $x = 2\pi$ and the y-axis at $y = 2$.

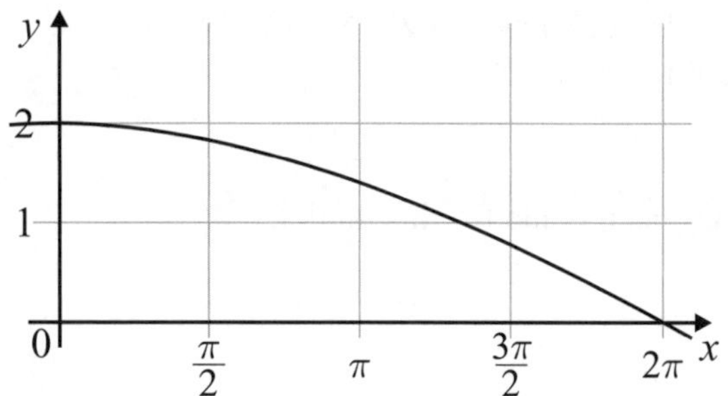

a) Lily works out $\int_0^{2\pi} 2\cos\frac{x}{4}\,dx$ and gets an answer of 4π. Her teacher doesn't do the integration, but tells her that she is wrong, and that her answer is too big. Explain how the teacher knew this by looking at the graph.

(1 mark)

b) Find the value of $\int_0^{2\pi} 2\cos\frac{x}{4}\,dx$.

(3 marks)

2 The diagram below shows a rectangular field 20 m long and 8 m wide enclosed by fences.
Inside the field is a rabbit pen, which is a regular hexagon with sides of length 2 m.
The pen is positioned with one of its sides up against one of the 20 m fences.

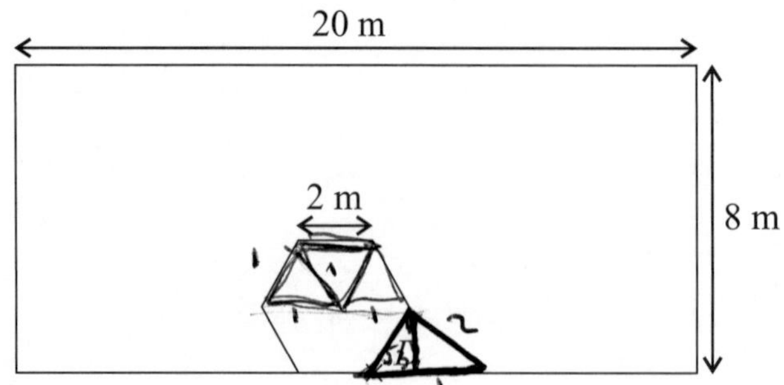

A dog is tied to the point labelled D on the diagram. The dog's lead is 2 m long. It is not able to enter the pen.

a) Calculate the exact area of the field the dog can reach.

(3 marks)

b) A cat now enters the field. It stays out of the area that the dog can get to and it is not able to enter the rabbit pen. Find the area of the field that the cat might go in. Give your answer correct to 3 significant figures.

(4 marks)

3 A group of musicians follow a 30-day programme to practise for a concert.

The length of time for which the pianist practises will increase by 10 minutes every day.
On day 1 the pianist practises for 1 hour.

a) On which day will the pianist practise for exactly 4 hours 40 minutes?

(3 marks)

The length of time for which the violinist practises will increase by 4% each day.
On day 1 the violinist practises for 1 hour 40 minutes.

b) On which day does the violinist first practise for more than 4 hours 40 minutes?

(4 marks)

c) Which of the two musicians practises for the longest time in total over the 30 days?
Give working to show how you came to your conclusion.

(3 marks)

4 The curve C has parametric equations

$$x = 1 + 5\cos t,\ \ y = 2 + 5\sin t,\quad 0 \le t \le 2\pi$$

a) Show that the Cartesian equation of the curve C is the circle given by the equation

$$(x - 1)^2 + (y - 2)^2 = 25$$

(3 marks)

The line l is the tangent to the circle at the point P (4, 6).

b) Find the equation of the line l. Give your answer in the form $ax + by + c = 0$, where a, b and c are integers to be found.

(4 marks)

5 **a)** Find $\frac{dy}{dx}$ for $y = 4x^3 + e^{x^2}$.

(3 marks)

b) Find the range of values of x for which y is concave downwards.

(5 marks)

6 **a)** Express $\frac{11x - 7}{(2x - 4)(x + 1)}$ in partial fractions.

(4 marks)

b) Hence find the binomial expansion of $\frac{11x - 7}{(2x - 4)(x + 1)}$, in ascending powers of x, up to and including the term in x^2.

(6 marks)

7 **In this question you must show detailed reasoning.**

Solve the equation $2\ \text{cosec}^2 x + 5 \cot x = 9$, for $-90° < x < 90°$.
Give your answers to 1 decimal place where appropriate.

(5 marks)

8 Part of a concrete sculpture is modelled as a cube with sides of length x m. Concrete costs £65 per cubic metre. Each face is then covered in gold leaf which costs £150 per square metre.

a) Given that the cost of materials used for the cube is £250, prove that x satisfies the equation $x = \sqrt{\frac{250 - 65x^3}{900}}$

(4 marks)

b) Hence show that the length of the cube is between 0.5 m and 1 m.

(2 marks)

c) Using the iteration formula $x_{n+1} = \sqrt{\frac{250 - 65x_n^3}{900}}$ with $x_1 = 0.75$, find x_2 and x_5. Give your answers correct to 4 decimal places.

(2 marks)

The graph below shows the curve $y = \sqrt{\frac{250 - 65x^3}{900}}$ and the line $y = x$. The position of x_1 is indicated.

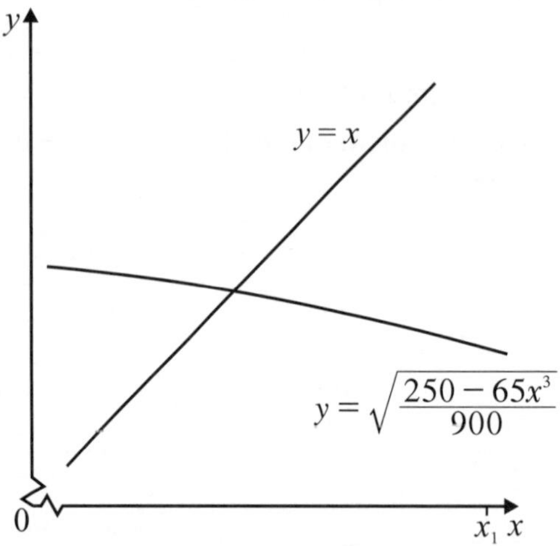

d) Use the graph to show the positions of x_2 and x_3.

(1 mark)

Section B (15 marks)

Pythagorean Triples

Pythagoras was born in Greece in approximately 570 BC. He is usually thought of as a mathematician, scientist and philosopher, although little is known for sure about his life.

Pythagoras famously gave his name to the theorem stating that, in any right-angled triangle, the square of the length of the hypotenuse, h, is equal to the sum of the squares of the other two side lengths, a and b.

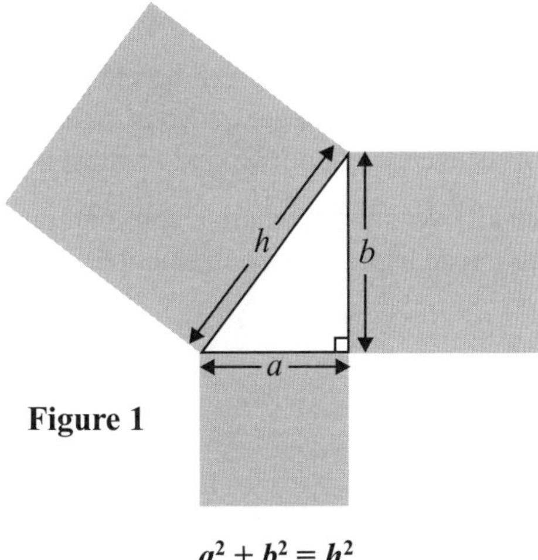

Figure 1

$$a^2 + b^2 = h^2$$

Even when the lengths of the two shorter sides (also known as legs) of a right-angled triangle have integer values, the length of the third side, the hypotenuse, is usually not a whole number. However, there are exceptions to this, and these exceptions are known as **Pythagorean triples**. For example, the numbers 3, 4 and 5 form a Pythagorean triple, since all are integers and the squares of the two smaller numbers added together equal the square of the larger number ($3^2 + 4^2 = 5^2$).

A Pythagorean triple such that the highest common factor of a, b and h is 1 is referred to as **primitive**.

There are various ways to generate Pythagorean triples. The simplest is to take a known Pythagorean triple and multiply each number by any whole number. A Pythagorean triple generated in this way is not primitive.

Another method for finding Pythagorean triples involves first rearranging the formula $a^2 + b^2 = h^2$ to give $a^2 = h^2 - b^2$. As a, b and h must be whole numbers, a^2 must be both a square number and the difference between two other square numbers. To simplify the search, only triples where h^2 and b^2 are consecutive squares will be considered. The differences between two consecutive square numbers form a sequence of odd numbers:

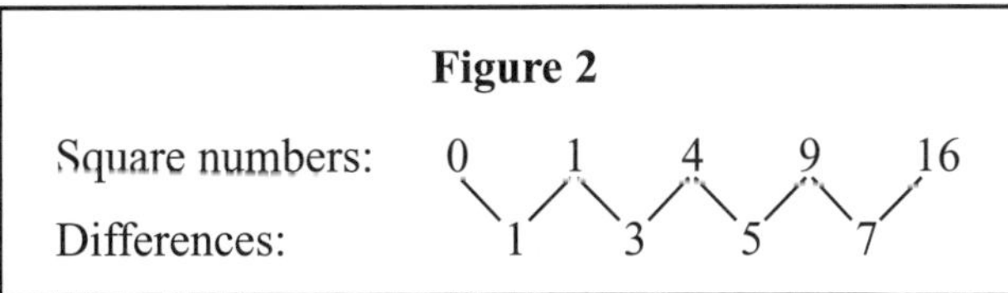

To generalise this in mathematical terms, the difference between $(n - 1)^2$ and n^2 is $2n - 1$.

A Pythagorean triple will exist when the difference between two consecutive square numbers is also a square number. For example, consider $a^2 = h^2 - b^2$ and suppose $a^2 = 9$. As 9 is odd, it can be expressed as the difference between two consecutive squares, and is also a square number itself. Continuing the sequence in Figure 2, the two consecutive square numbers which have a difference of 9 are 16 and 25, which correspond to a^2, b^2 and h^2. Taking the square roots of these numbers gives the primitive Pythagorean triple 3, 4, 5.

Continued on next page

A third method for finding a Pythagorean triple is to take any two positive integers with a difference of 2. Create two unit fractions (fractions whose numerator is 1 and denominator is a positive integer) with these numbers as denominators, and add them together. The numerator and the denominator of the result will give the two legs of a Pythagorean triple.

The legs in a primitive Pythagorean triple cannot both be even. This can be proved as follows:

- Suppose a, b and h form a primitive Pythagorean triple with $a^2 + b^2 = h^2$, and that a and b are both even.
- Since a is even, a can be written as $a = 2m$ for some integer m. So $a^2 = (2m)^2 = 4m^2$ is also even. Similarly, since b is even, b^2 must also be even.
- As $a^2 + b^2 = h^2$, h^2 must be even too, since the sum of two even numbers is always even.
- The square of an odd number is always odd. This can be seen by considering an odd integer p. Since p is odd, it can be written $p = 2q + 1$, for some integer q. So $p^2 = (2q + 1)^2 = 4q^2 + 4q + 1 = 4(q^2 + q) + 1$, which is odd. Therefore, as h^2 is even, h cannot be odd so must be even.
- This means that a, b and h are all even, and so share a common factor of 2. In other words, a, b and h do not form a primitive Pythagorean triple — a contradiction. Hence a and b cannot both be even.

9 Prove algebraically that, when the three numbers of a Pythagorean triple are all multiplied by the same positive integer, the result is also a Pythagorean triple.

(2 marks)

10 Using $a^2 = 49$, find another Pythagorean triple using the method described in lines 13-23.

(2 marks)

11 The Pythagorean triple 9, 40, 41 was generated using the method described in lines 24-27. Find the positive integers used as the denominators in the unit fractions.

(3 marks)

12 **a)** The method described in lines 24-27 starts with two numbers. By writing the smaller of these numbers as $n - 1$, prove that the numerator and the denominator of the sum of the two fractions always form two legs of a Pythagorean triple.

(3 marks)

b) If x and y are the two legs in the triple described in part a), where x is even and y is odd, express y in terms of x.

(1 mark)

13 The positive integers a, b and h form a primitive Pythagorean triple.

a) Show that for any even integer d, the number d^2 is always divisible by 4.

(1 mark)

b) Hence or otherwise prove that a and b cannot both be odd.

(3 marks)

END OF EXAM PAPER

TOTAL FOR PAPER: 75 MARKS

Section One — Pure Maths

Pages 3-4: Proof

1 Proof by exhaustion: if n is even, n^3 is also even (as the product of even numbers is even), so $n^3 - n$ is even too (the difference between two even numbers is always even) ***[1 mark]***.
If n is odd, n^3 is also odd (as the product of odd numbers is odd), so $n^3 - n$ is even (as an odd number minus an odd number is even) ***[1 mark]***. n is an integer so must be odd or even, so $n^3 - n$ is always even.
You could have factorised $n^3 - n$ instead — you'd get $n(n + 1)(n - 1)$. Then if n is odd, $(n + 1)$ and $(n - 1)$ are even, so the product is even. If n is even, the product will be even too.

2 Proof by contradiction: Assume that there is a largest integer, n ***[1 mark]***. Now consider a number $k = n + 1$. k is an integer, as the sum of two integers is itself an integer, and $k = n + 1 > n$ ***[1 mark]***. This contradicts the assumption that n is the largest integer ***[1 mark]***. Hence, there is no largest integer.

3 Take two prime numbers, p and q ($p \neq q$ and $p, q > 1$). As p is prime, its only factors are 1 and p ***[1 mark]***, and as q is prime, its only factors are 1 and q ***[1 mark]***. So the product pq has factors 1, p, q, and pq ***[1 mark]*** (these factors are found by multiplying the factors of each number together in every possible combination). $pq \neq 1$ as $p, q > 1$. Hence the product of any two distinct prime numbers has exactly four factors.

4 As a and b are rational, you can write $a = \frac{m}{n}$ and $b = \frac{p}{q}$ where m, n, p and q are integers ***[1 mark]***.
So, $c = \frac{m}{n} - \frac{p}{q} = \frac{mq - pn}{nq}$ ***[1 mark]***.
The product of two integers is always an integer and the difference of two integers is also always an integer.
Hence, $c = \frac{mq - pn}{nq}$ is also a rational number. ***[1 mark]***

5 As n is even, write n as $2k$ for some integer k ***[1 mark]***.
So $n^3 + 2n^2 + 12n = (2k)^3 + 2(2k)^2 + 12(2k)$
$= 8k^3 + 8k^2 + 24k$ ***[1 mark]***
This can be written as $8x$, where $x = k^3 + k^2 + 3k$, so is always a multiple of 8 when n is even ***[1 mark]***.

6 a) E.g. The student's proof shows that if x is even, then x^3 is even. This is not the same as showing if x^3 is even, then x is even.
[1 mark for any sensible explanation of why the proof is not valid]

b) Proof by contradiction:
Assume that the statement is false. So, that means for some value of x, x^3 is even but x is odd ***[1 mark]***.
x is odd so $x = 2n + 1$, where n is an integer ***[1 mark]***.
So, $x^3 = (2n + 1)^3 = (4n^2 + 4n + 1)(2n + 1)$
$= 8n^3 + 12n^2 + 6n + 1$
$= 2(4n^3 + 6n^2 + 3n) + 1$ ***[1 mark]***
$4n^3 + 6n^2 + 3n$ is an integer, call it m, so $x^3 = 2m + 1$ is odd.
This contradicts the assumption that x^3 is even but x is odd, ***[1 mark]*** hence if x^3 is even, then x is even.

7 a) E.g. Let both x and y be $\sqrt{2}$. $\sqrt{2}$ is irrational, but $\frac{\sqrt{2}}{\sqrt{2}} = 1$ which is rational, hence Riyad's claim is false.
[1 mark for any valid counter-example]

b) Let x be a rational number, where $x \neq 0$.
Let z be an irrational number.
Assume that the product zx is rational, so $zx = \frac{a}{b}$ where a and b are integers ($a, b \neq 0$) ***[1 mark]***.
x is rational, so $x = \frac{l}{m}$ where l and m are integers ($l, m \neq 0$).
So, the product $zx = \frac{zl}{m} = \frac{a}{b} \Rightarrow z = \frac{ma}{lb}$ ***[1 mark]***.
ma and lb are the products of non-zero integers, so are non-zero integers themselves. So, z is rational. This is a contradiction, as z is irrational by definition, ***[1 mark]*** therefore, zx must be irrational.

Pages 5-7: Algebra and Functions 1

1 $\sqrt[m]{a^n} = a^{\frac{n}{m}}$ so $\sqrt{a^4} = a^2$ and $a^6 \times a^3 = a^{6+3} = a^9$
So $\frac{a^6 \times a^3}{\sqrt{a^4}} \div a^{\frac{1}{2}} = \frac{a^9}{a^2} \div a^{\frac{1}{2}}$ ***[1 mark]***
$= a^{7 - \frac{1}{2}} = a^{\frac{13}{2}}$ ***[1 mark]***

2 a) $3 = \sqrt[3]{27} = 27^{\frac{1}{3}}$, so $x = \frac{1}{3}$ ***[1 mark]***
b) $81 = 3^4 = (\sqrt[3]{27})^4 = 27^{\frac{4}{3}}$, so $x = \frac{4}{3}$ ***[1 mark]***

3 $\frac{(3ab^3)^2 \times 2a^6}{6a^4b} = \frac{3^2 \times a^2 \times (b^3)^2 \times 2a^6}{6a^4b} = \frac{18a^8b^6}{6a^4b} = 3a^4b^5$
[2 marks available — 1 mark for simplifying the numerator to get $18a^8b^6$, 1 mark for the correct answer]

4 $\frac{(5 + 4\sqrt{x})^2}{2x} = \frac{25 + 40\sqrt{x} + 16x}{2x}$ ***[1 mark]***
$= \frac{1}{2}x^{-1}(25 + 40x^{\frac{1}{2}} + 16x)$ ***[1 mark]***
$= \frac{25}{2}x^{-1} + 20x^{-\frac{1}{2}} + 8$ ***[1 mark]***
So P = 20 and Q = 8.

5 $(5\sqrt{5} + 2\sqrt{3})^2 = (5\sqrt{5} + 2\sqrt{3})(5\sqrt{5} + 2\sqrt{3})$
$= (5\sqrt{5})^2 + 2(5\sqrt{5} \times 2\sqrt{3}) + (2\sqrt{3})^2$
First term: $(5\sqrt{5})^2 = 5\sqrt{5} \times 5\sqrt{5}$
$= 5 \times 5 \times \sqrt{5} \times \sqrt{5} = 5 \times 5 \times 5 = 125$ ***[1 mark]***
Second term: $2(5\sqrt{5} \times 2\sqrt{3}) = 2 \times 5 \times 2 \times \sqrt{5} \times \sqrt{3}$
$= 20\sqrt{15}$ ***[1 mark]***
Third term: $(2\sqrt{3})^2 = 2\sqrt{3} \times 2\sqrt{3} = 2 \times 2 \times \sqrt{3} \times \sqrt{3}$
$= 2 \times 2 \times 3 = 12$ ***[1 mark]***
So $(5\sqrt{5})^2 + 2(5\sqrt{5} \times 2\sqrt{3}) + (2\sqrt{3})^2$
$= 125 + 20\sqrt{15} + 12 = 137 + 20\sqrt{15}$
So $a = 137$, $b = 20$ and $c = 15$ ***[1 mark]***.

6 Multiply top and bottom by $\sqrt{5} - 1$:
$\frac{10 \times (\sqrt{5} - 1)}{(\sqrt{5} + 1) \times (\sqrt{5} - 1)}$ ***[1 mark]***
$= \frac{10(\sqrt{5} - 1)}{5 - 1} = \frac{10\sqrt{5} - 10}{4} = \frac{5\sqrt{5} - 5}{2}\left(= \frac{5(\sqrt{5} - 1)}{2}\right)$
[1 mark for simplifying top, 1 mark for simplifying bottom]

7 Rationalise the denominator by multiplying top and bottom by $(2 - \sqrt{2})$:
$\frac{4 + \sqrt{2}}{2 + \sqrt{2}} \times \frac{2 - \sqrt{2}}{2 - \sqrt{2}} = \frac{8 - 4\sqrt{2} + 2\sqrt{2} - 2}{4 - 2}$
$= \frac{6 - 2\sqrt{2}}{2} = 3 - \sqrt{2}$
[3 marks available — 1 mark for multiplying numerator and denominator by the correct expression, 1 mark for correct multiplication, 1 mark for the correct answer]

8 $\frac{(x^2 - 9)(3x^2 - 10x - 8)}{(6x + 4)(x^2 - 7x + 12)} = \frac{(x + 3)(x - 3)(3x + 2)(x - 4)}{2(3x + 2)(x - 3)(x - 4)}$
$= \frac{x + 3}{2}$
[3 marks available — 1 mark for factorising the numerator, 1 mark for factorising the denominator and 1 mark for the correct answer]

9 a) $\frac{x^2 + 5x - 14}{2x^2 - 4x} = \frac{(x + 7)(x - 2)}{2x(x - 2)} = \frac{x + 7}{2x}$
[2 marks available — 1 mark for factorising the numerator and the denominator and 1 mark for cancelling to obtain correct answer]

b) $\dfrac{x^2+5x-14}{2x^2-4x}+\dfrac{14}{x(x-4)}=\dfrac{x+7}{2x}+\dfrac{14}{x(x-4)}$

$=\dfrac{(x+7)(x-4)}{2x(x-4)}+\dfrac{2\times 14}{2x(x-4)}$

$=\dfrac{x^2+3x-28+28}{2x(x-4)}=\dfrac{x^2+3x}{2x(x-4)}$

$=\dfrac{x(x+3)}{2x(x-4)}=\dfrac{x+3}{2(x-4)}$

[3 marks available — 1 mark for putting fractions over a common denominator, 1 mark for multiplying out and simplifying the numerator and 1 mark for cancelling to obtain correct answer]

10 $\dfrac{1}{x(2x-3)}\equiv\dfrac{A}{x}+\dfrac{B}{2x-3}\Rightarrow 1\equiv A(2x-3)+Bx$ ***[1 mark]***

Equating constants gives $1=-3A\Rightarrow A=-\frac{1}{3}$

Equating coefficients of x gives $0=2A+B\Rightarrow B=\frac{2}{3}$

[1 mark for A or B]

So $\dfrac{1}{x(2x-3)}\equiv\dfrac{2}{3(2x-3)}-\dfrac{1}{3x}$ ***[1 mark]***

11 $\dfrac{6x-1}{x^2+4x+4}\equiv\dfrac{6x-1}{(x+2)^2}$ ***[1 mark]*** $\equiv\dfrac{A}{(x+2)}+\dfrac{B}{(x+2)^2}$

$\Rightarrow 6x-1\equiv A(x+2)+B$ ***[1 mark]***

Equating coefficients of x gives $6=A$

Equating constant terms gives $-1=2A+B\Rightarrow B=-13$

[1 mark for A or B]

So $\dfrac{6x-1}{x^2+4x+4}\equiv\dfrac{6}{(x+2)}-\dfrac{13}{(x+2)^2}$ ***[1 mark]***

Pages 8-11: Algebra and Functions 2

1 When $ax^2+bx+c=0$ has no real roots, you know that $b^2-4ac<0$ ***[1 mark]***. Here, $a=-j$, $b=3j$ and $c=1$. Therefore $(3j)^2-(4\times -j\times 1)<0\Rightarrow 9j^2+4j<0$ ***[1 mark]***

To find the values where $9j^2+4j<0$, you need to start by solving $9j^2+4j=0$: $j(9j+4)=0$, so $j=0$ or $9j=-4\Rightarrow j=-\frac{4}{9}$

Now sketch the graph:

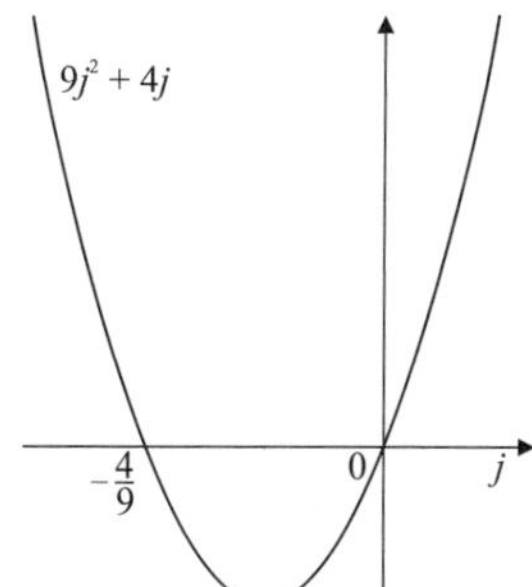

From the graph, you can see that $9j^2+4j<0$ when $-\frac{4}{9}<j<0$ ***[1 mark]***.

2 a) Complete the square by halving the coefficient of x to find the number in the brackets (m):

$x^2-7x+17=\left(x-\frac{7}{2}\right)^2+n$

$\left(x-\frac{7}{2}\right)^2=x^2-7x+\frac{49}{4}$, so $n=17-\frac{49}{4}=\frac{19}{4}$

So $x^2-7x+17=\left(x-\frac{7}{2}\right)^2+\frac{19}{4}$

[3 marks available — 1 mark for the correct brackets, 1 mark for finding the right correcting number and 1 mark for the correct final answer]

b) The maximum value of f(x) will be when the denominator is as small as possible — so you want the minimum value of $x^2-7x+17$. Using the completed square above, you can see that the minimum value is $\frac{19}{4}$ ***[1 mark]*** because the squared part can equal but never be below 0.

So the maximum value of f(x) is $\dfrac{1}{\left(\frac{19}{4}\right)}=\dfrac{4}{19}$ ***[1 mark]***.

3 If the equation has two real roots, then $b^2-4ac>0$ ***[1 mark]***.

For this equation, $a=3k$, $b=k$ and $c=2$.

Use the discriminant formula to find k:

$k^2-(4\times 3k\times 2)>0$ ***[1 mark]***

$k^2-24k>0$

$k(k-24)>0$

$k<0$ or $k>24$ ***[1 mark]***

If you're struggling to solve this inequality, you could always sketch a graph like in the answer to question 1.

4 Let $y=x^3$. Then $x^6=7x^3+8$ becomes $y^2=7y+8$ ***[1 mark]***, so solve the quadratic in y:

$y^2=7y+8\Rightarrow y^2-7y-8=0$

$(y-8)(y+1)=0$, so $y=8$ or $y=-1$ ***[1 mark]***.

Now replace y with x^3. So $x^3=8\Rightarrow x=2$ ***[1 mark]*** or $x^3=-1\Rightarrow x=-1$ ***[1 mark]***.

Here, you had to spot that the original equation was a quadratic of the form x^2+bx+c, just in terms of x^3 not x.

5 a) (i) First, rewrite the quadratic as: $-h^2+10h-27$ and complete the square ($a=-1$):

$-(h-5)^2+25-27=-(h-5)^2-2$

Rewrite the square in the form given in the question:

$T=-(-(5-h))^2-2\Rightarrow T=-(5-h)^2-2$

[3 marks available — 1 mark for $(5-h)^2$ or $(h-5)^2$, 1 mark for 25 – 27, 1 mark for the correct final answer]

The last couple of steps are using the fact that $(-a)^2=a^2$ to show that $(m-n)^2=(n-m)^2$...

(ii) $(5-h)^2\geq 0$ for all values of h, so $-(5-h)^2\leq 0$. Therefore $-(5-h)^2-2<0$ for all h, so T is always negative ***[1 mark]***.

b) (i) The maximum temperature is the maximum value of T, which is -2 (from part a) ***[1 mark]***, and this occurs when the expression in the brackets $=0$. The h-value that makes the expression in the brackets 0 is 5 ***[1 mark]***, so maximum temperature occurs 5 hours after sunrise.

(ii) At sunrise, $h=0$, so $T=10(0)-0^2-27=-27$°C, so the graph looks like this:

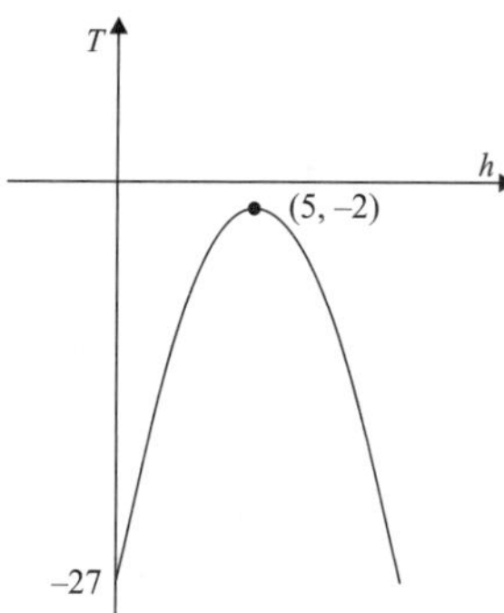

[2 marks available — 1 mark for drawing n-shaped curve that sits below the x-axis with the maximum roughly where shown (even if its position is not labelled), 1 mark for correct T-axis intercept (0, –27)]

6 Rearrange the first equation to get y on its own:

$y+x=7\Rightarrow y=7-x$ ***[1 mark]***

Substitute the expression for y into the quadratic to get:

$7-x=x^2+3x-5$ ***[1 mark]***

Rearrange again to get everything on one side of the equation, and then factorise it:

$0=x^2+4x-12\Rightarrow (x+6)(x-2)=0$

So $x=-6$ and $x=2$ ***[1 mark]***

Use these values to find the corresponding values of y:

When $x=-6$, $y=7--6=13$

and when $x=2$, $y=7-2=5$ ***[1 mark for both y-values]***

So the solutions are $x=-6$, $y=13$ or $x=2$, $y=5$.

7 a) At points of intersection, $-2x+4=-x^2+3$ ***[1 mark]***

$x^2-2x+1=0$

$(x-1)^2=0$ so $x=1$ ***[1 mark]***.

When $x=1$, $y=-2x+4=2$,

so there is one point of intersection at $(1, 2)$ ***[1 mark]***.

b)

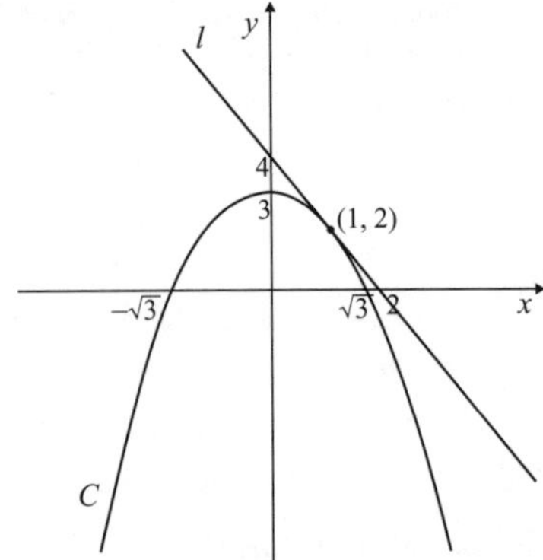

[5 marks available — 1 mark for drawing n-shaped curve, 1 mark for x-axis intercepts at $\pm\sqrt{3}$, 1 mark for maximum point of curve and y-axis intercept at (0, 3). 1 mark for line crossing the y-axis at (0, 4) and the x-axis at (2, 0). 1 mark for line and curve touching in one place at (1, 2).]

8 Draw the line $y = x + 2$, which has a gradient of 1, crosses the y-axis at (0, 2) and crosses the x-axis at (–2, 0). This should be a solid line.
Then draw the curve $y = 4 - x^2 = (2 + x)(2 - x)$. This is an n-shaped quadratic which crosses the x-axis at (–2, 0) and (2, 0) and the y-axis at (0, 4) (this is also the maximum point of the graph). This should be a dotted line.
Then test the point (0, 0) to see which side of the lines you want:
$0 \geq 0 + 2$ — this is false, so shade the other side of the line.
$4 - 0^2 > 0$ — this is true, so shade the region below the curve.
So the final region (labelled R) should look like this:

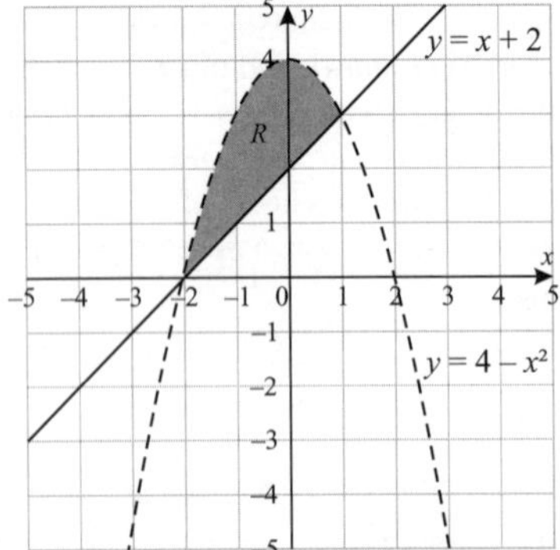

[3 marks available — 1 mark for drawing the line with correct gradient and intercepts, 1 mark for drawing the curve with correct intercepts, 1 mark for shading or indicating the correct region]

9 $x^2 - 8x + 15 > 0 \Rightarrow (x - 5)(x - 3) > 0$
Sketch a graph to see where the quadratic is greater than 0 it'll be a u-shaped curve that crosses the x-axis at $x = 3$ and $x = 5$.

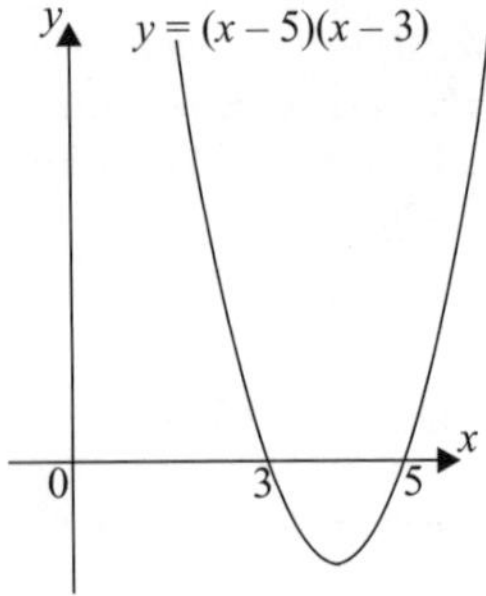

You can see from the graph that the function is positive when $x < 3$ and when $x > 5$. In set notation, this is $\{x : x < 3\} \cup \{x : x > 5\}$.
[4 marks available — 1 mark for factorising the quadratic, 1 mark for finding the roots, 1 mark for x < 3 and x > 5, 1 mark for the correct answer in set notation]

10 a) Multiply out the brackets and rearrange to get 0 on one side:
$(x - 1)(x^2 + x + 1) = 2x^2 - 17$
$x^3 + x^2 + x - x^2 - x - 1 = 2x^2 - 17$ ***[1 mark]***
$x^3 - 2x^2 + 16 = 0$ ***[1 mark]***

b) By the factor theorem, if $(x + 2)$ is a factor of f(x), f(–2) = 0.
$f(x) = x^3 - 2x^2 + 16$
$f(-2) = (-2)^3 - 2(-2)^2 + 16 = -8 - 8 + 16 = 0$ ***[1 mark]***
f(–2) = 0, therefore $(x + 2)$ is a factor of f(x) ***[1 mark]***.

c) From part b) you know that $(x + 2)$ is a factor of f(x).
Dividing f(x) by $(x + 2)$ gives:
$x^3 - 2x^2 + 16 = (x + 2)(x^2 + ?x + 8) = (x + 2)(x^2 - 4x + 8)$
If you find it easier, you can use algebraic long division here.
[2 marks available — 2 marks for all three correct terms in the quadratic, otherwise 1 mark for two terms correct]

d) From b) you know that $x = -2$ is a root. From c),
$f(x) = (x + 2)(x^2 - 4x + 8)$. So for f($x$) to equal zero,
either $(x + 2) = 0$ (so $x = -2$) or $(x^2 - 4x + 8) = 0$ ***[1 mark]***.
Completing the square of $(x^2 - 4x + 8)$ gives
$x^2 - 4x + 8 = (x - 2)^2 + 4$, which is always positive so has no real roots. So f(x) = 0 has no solutions other than $x = -2$, which means it only has one root ***[1 mark]***.
You could also have shown that $x^2 - 4x + 8$ has no real roots by showing that the discriminant is < 0.

11 If $(x - 1)$ is a factor of f(x), then f(1) = 0 by the factor theorem ***[1 mark]***.
$f(1) = 1^3 - 4(1)^2 - a(1) + 10$, so $0 = 7 - a \Rightarrow a = 7$ ***[1 mark]***.
So $f(x) = x^3 - 4x^2 - 7x + 10$.
To solve f(x) = 0, first factorise $x^3 - 4x^2 - 7x + 10$. You know one factor, $(x - 1)$, so find the quadratic that multiplies with that factor to give the original equation:
$x^3 - 4x^2 - 7x + 10 = (x - 1)(x^2 + ?x - 10)$
$= (x - 1)(x^2 - 3x - 10)$ ***[1 mark]***
Again, you could use algebraic long division to do this.
Then factorise the quadratic: $= (x - 1)(x - 5)(x + 2)$ ***[1 mark]***
Finally, solve f(x) = 0:
$x^3 - 4x^2 - 7x + 10 = 0 \Rightarrow (x - 1)(x - 5)(x + 2) = 0$, so $x = 1$, $x = 5$ or $x = -2$ ***[2 marks for all three x-values, otherwise 1 mark for either x = 5 or x = –2]***.

Pages 12-17: Algebra and Functions 3

1 a)

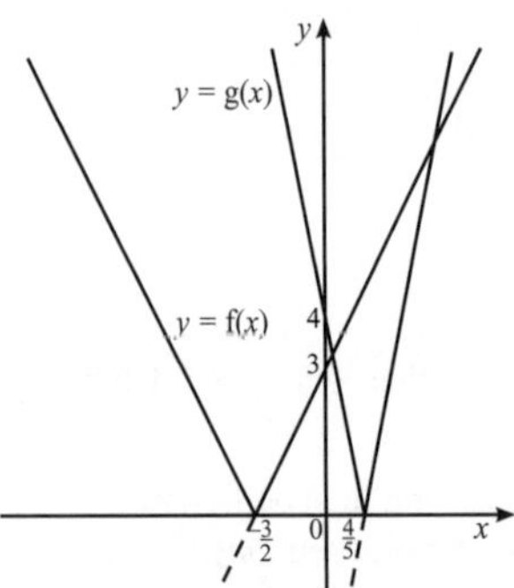

[2 marks available — 1 mark for y = |2x + 3| (with correct x- and y-intercepts), 1 mark for y = |5x − 4| (with correct x- and y-intercepts)]

b) From the graph, it is clear that there are two points where the graphs intersect. One is in the range $-\frac{3}{2} < x < \frac{4}{5}$, where $(2x + 3) > 0$ but $(5x - 4) < 0$.
This gives $2x + 3 = -(5x - 4)$ ***[1 mark]***.
The other one is in the range $x > \frac{4}{5}$, where $(2x + 3) > 0$ and $(5x - 4) > 0$, so $2x + 3 = 5x - 4$ ***[1 mark]***. Solving the first equation gives: $2x + 3 = -5x + 4 \Rightarrow 7x = 1$, so $x = \frac{1}{7}$ ***[1 mark]***. Solving the second equation gives:
$2x + 3 = 5x - 4 \Rightarrow 7 = 3x$, so $x = \frac{7}{3}$ ***[1 mark]***.

2 a) If $|x| = 2$, then either $x = 2$ or $x = -2$
When $x = 2$, $|4x + 5| = |8 + 5| = |13| = 13$ ***[1 mark]***
When $x = -2$, $|4x + 5| = |-8 + 5|$ ***[1 mark]*** $= |-3| = 3$ ***[1 mark]***

b) First, sketch a quick graph:

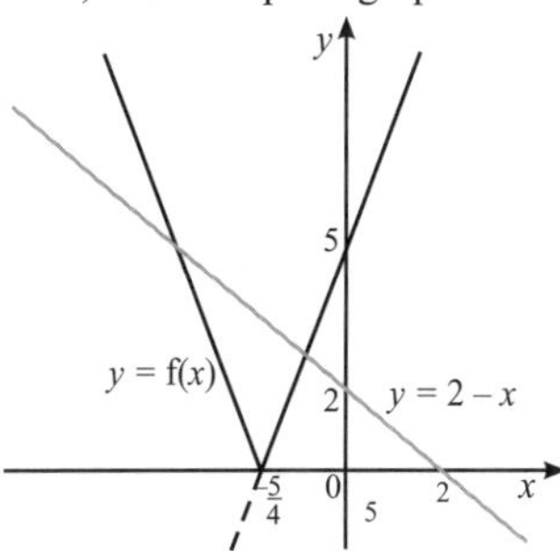

You can see that the lines cross twice, so you need to solve two inequalities:

$4x + 5 \leq 2 - x \Rightarrow 5x \leq -3 \Rightarrow x \leq -\frac{3}{5}$

and $-(4x + 5) \leq 2 - x \Rightarrow -7 \leq 3x \Rightarrow x \geq -\frac{7}{3}$

So $f(x) \leq 2 - x$ when $-\frac{7}{3} \leq x \leq -\frac{3}{5}$

[3 marks available — 1 mark for each correct value in the inequality, 1 mark for the correct inequality signs]

c) Two distinct roots means that the graphs of $y = f(x) + 2$ and $y = A$ cross twice ***[1 mark]***. From the graph in part b), the graph of $y = f(x) + 2$ is the black line translated up by 2. A horizontal line will intersect this in two places as long as it lies above the point where the graph is reflected, i.e. above $y = 2$. So the possible values of A are $A > 2$ ***[1 mark]***.

3 The quartic has already been factorised — there are two double roots, one at (2, 0) and the other at (−3, 0). When $x = 0$, $y = (-2)^2(3^2) = 36$, so the y-intercept is (0, 36). The coefficient of the x^4 term is positive, and as the graph only touches the x-axis but doesn't cross it, it is always above the x-axis. The graph looks like this:

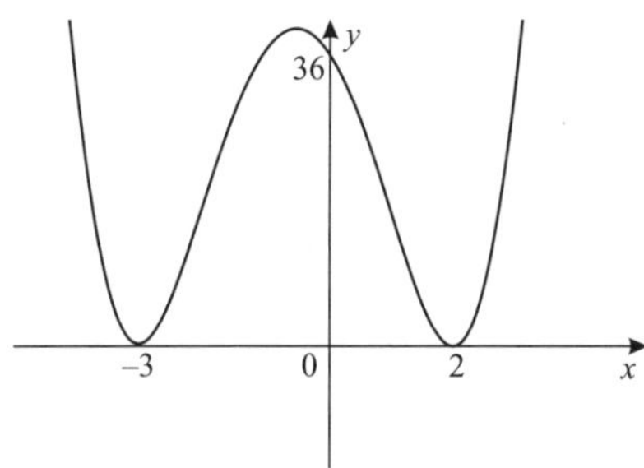

[3 marks available — 1 mark for the correct shape, 1 mark for the correct x-intercepts, 1 mark for the correct y-intercept]

4 a)

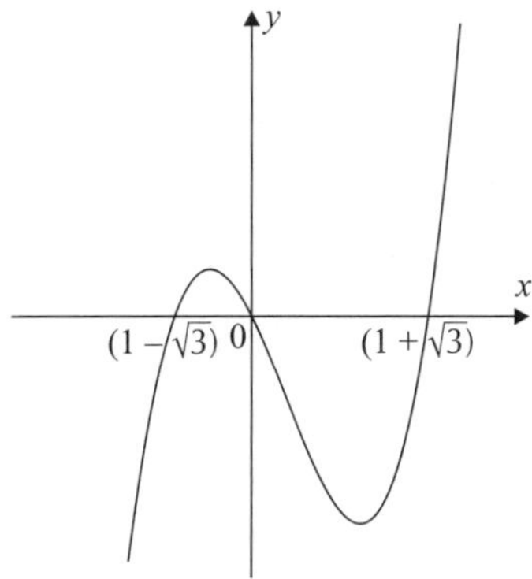

[2 marks available — 1 mark for the correct positive cubic shape with the two turning points the correct side of the y-axis and 1 mark for the x-intercepts correctly labelled]

b) $x^3 - 2x^2 + px$ can be factorised to give $x(x^2 - 2x + p)$, so the roots of the quadratic factor must be $x = 1 + \sqrt{3}$ and $x = 1 - \sqrt{3}$. The quadratic factor can be factorised to give $(x - (1 + \sqrt{3}))(x - (1 - \sqrt{3}))$, so the constant term is given by $p = (1 + \sqrt{3})(1 - \sqrt{3}) = 1 - 3 = -2$

[2 marks available — 1 mark for a correct method to find p, 1 mark for the correct answer]

5 To transform the curve $y = x^3$ into $y = (x - 1)^3$, translate it 1 unit horizontally to the right (in the positive x-direction) ***[1 mark]***. To transform this into the curve $y = 2(x - 1)^3$, stretch it vertically (parallel to the y-axis) by a scale factor of 2 ***[1 mark]***. Finally, to transform into the curve $y = 2(x - 1)^3 + 4$, the whole curve is translated 4 units upwards (in the positive y-direction) ***[1 mark]***.

6 a) Expand the brackets to show the two functions are the same:

$$(2t + 1)(t - 2)(t - 3.5) = (2t + 1)(t^2 - 5.5t + 7)$$
$$= 2t^3 - 11t^2 + 14t + t^2 - 5.5t + 7$$
$$= 2t^3 - 10t^2 + 8.5t + 7 \text{ as required}$$

[2 marks available — 1 mark for expanding the brackets, 1 mark for rearranging to get the required answer]

You could also have shown this by using the factor theorem and showing that when t = −0.5, t = 2 or t = 3.5 then V = 0. You'd still get all the marks for using this method correctly.

b)

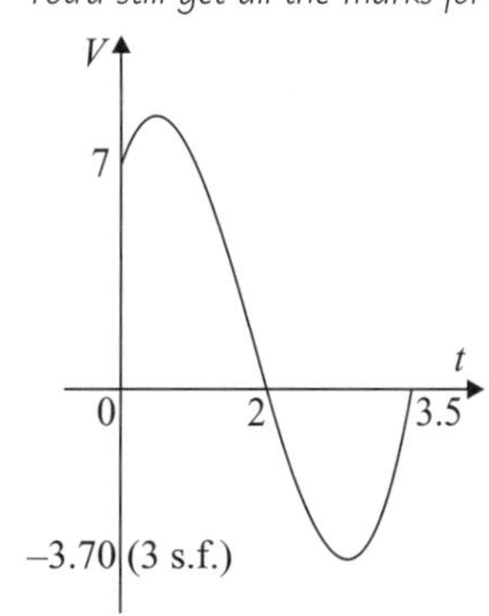

[3 marks available — 1 mark for the correct shape drawn between t = 0 and t = 3.5, 1 mark for the correct t-intercepts, 1 mark for the correct V-intercept]

c) 2 s ***[1 mark]***

This is the first point on the graph at which V = 0.

d) When the diver starts his dive (i.e. $t = 0$), $V = 7$. The diver's height is 1.75 m, so the diving board is 7 − 1.75 = 5.25 m high.

[2 marks available — 1 mark for a correct method, 1 mark for the correct answer]

e) The adapted model is a vertical translation of the original graph by 3 m upwards. This means that the lowest point of the dive is 3.70 − 3 = 0.70 m below the surface of the pool. This is obviously unrealistic as it is far too shallow — you would expect the diver to go at least as deep as from the lower diving board, so the adapted model is not valid.

[2 marks available — 1 mark for stating that the model is not valid, 1 mark for a sensible explanation of why]

This model is a lot easier to comment on if you realise it's just a vertical translation of the original model.

7 a)

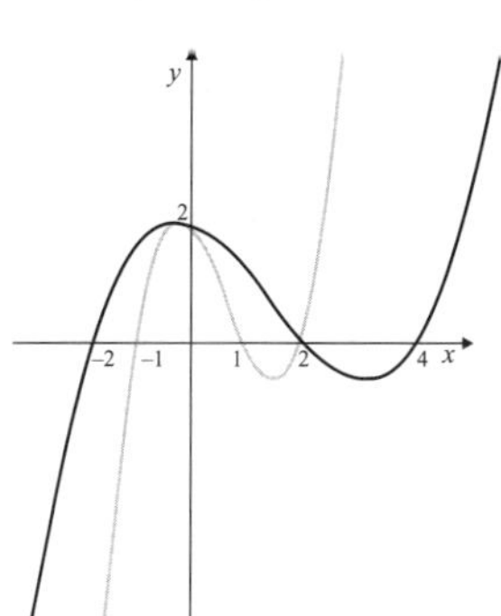

[3 marks available — 1 mark for horizontal stretch, 1 mark for x-axis intercepts at −2, 2 and 4, 1 mark for correct y-axis intercept at 2]

b) 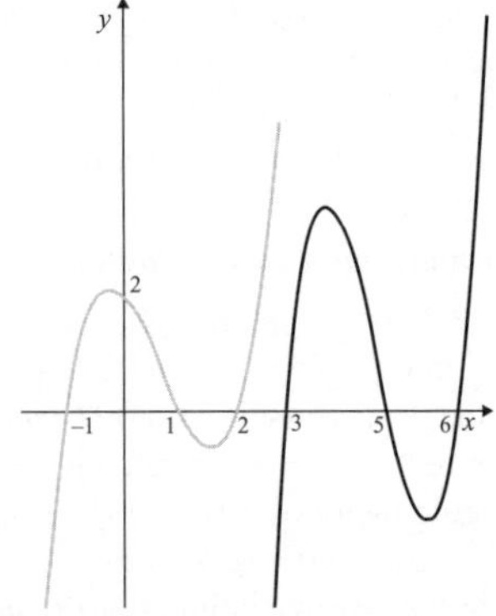

[3 marks available — 1 mark for vertical stretch, 1 mark for horizontal translation to the right, 1 mark for x-axis intercepts at 3, 5 and 6]

8 a) 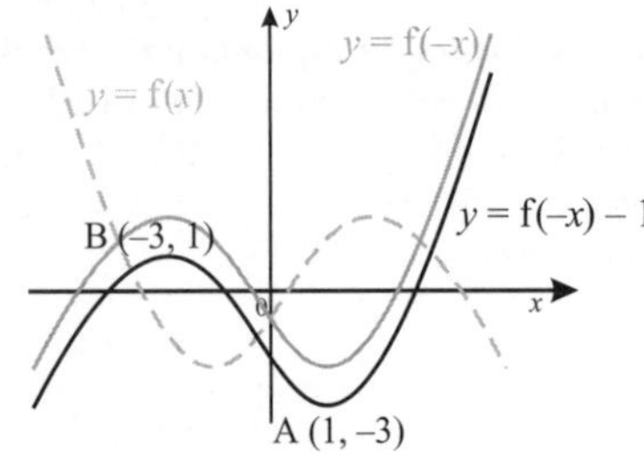

[3 marks available — 1 mark for shape (reflection in the y-axis and translation), 1 mark each for coordinates of A and B after transformation]

The solid grey line shows the graph of y = f(−x) — it's easier to do the transformation in two stages, instead of doing it all at once.

b) 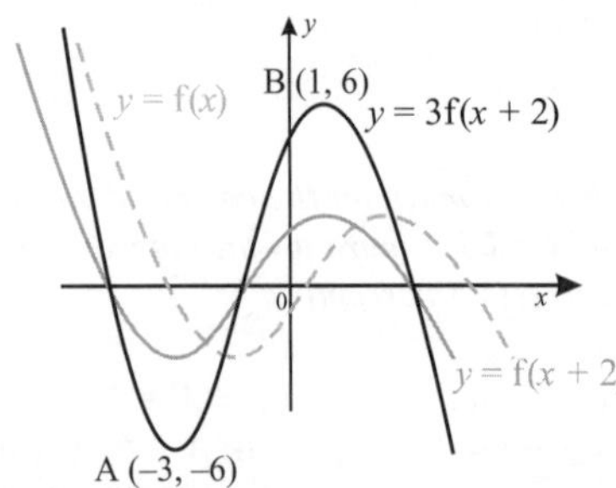

[3 marks available — 1 mark for shape (stretch and translation), 1 mark each for coordinates of A and B after transformation]

9 a) A translation of 3 up is a translation of the form $f(x) + 3$ ***[1 mark]***, then a translation of 2 right is $f(x-2) + 3$ ***[1 mark]***. Finally, a reflection in the y-axis gives $f(-(x-2)) + 3$. So $g(x) = f(2-x) + 3$ ***[1 mark]***.

b) Original coordinates of P: (1, 2).
After translation of 3 up and 2 right: (3, 5)
After reflection in the y-axis: (−3, 5) ***[1 mark]***.
Original coordinates of Q: (3, 16).
After translation of 3 up and 2 right: (5, 19)
After reflection in the y-axis: (−5, 19) ***[1 mark]***.
Do a quick sketch of the graph if you need to.

10 a) (i) $gf(x) = g(2^x) = \sqrt{3(2^x)+1}$ ***[1 mark]***

(ii) $gf(x) = 5 \Rightarrow \sqrt{3(2^x)+1} = 5 \Rightarrow 3(2^x) + 1 = 25$
$\Rightarrow 3(2^x) = 24 \Rightarrow 2^x = 8 \Rightarrow x = 3$
[2 marks available — 1 mark for a correct method, 1 mark for the correct answer]

b) First write $y = g(x)$ and rearrange to make x the subject:
$y = \sqrt{3x+1} \Rightarrow y^2 = 3x + 1 \Rightarrow y^2 - 1 = 3x$
$\Rightarrow \dfrac{y^2-1}{3} = x$
Then replace x with $g^{-1}(x)$ and y with x:
$g^{-1}(x) = \dfrac{x^2-1}{3}$ ***[1 mark]***
$g(x)$ has domain $x \geq -\frac{1}{3}$ and range $g(x) \geq 0$, so $g^{-1}(x)$ has domain $x \geq 0$ ***[1 mark]*** and range $g^{-1}(x) \geq -\frac{1}{3}$ ***[1 mark]***.

11 a) g has range $g(x) \geq -k$ ***[1 mark]***, as the minimum value of g is $-k$.

b) Neither f nor g are one-to-one functions, so they don't have inverses ***[1 mark]***.

c) (i) $gf(1) = g\left(\frac{1}{1^2}\right) = g(1) = 1^2 - k$ ***[1 mark]***
So $1 - k = -8 \Rightarrow k = 9$ ***[1 mark]***
$fg(x) = f(x^2 - 9)$ ***[1 mark]*** $= \dfrac{1}{(x^2-9)^2}$ ***[1 mark]***.
The domain of fg is $x \in \mathbb{R}, x \neq \pm 3$ ***[1 mark]***, as the denominator of the function can't be 0.

(ii) From part (i), you know that $fg(x) = \dfrac{1}{(x^2-9)^2}$, so
$\dfrac{1}{(x^2-9)^2} = \dfrac{1}{256} \Rightarrow (x^2-9)^2 = 256$ ***[1 mark]***
$x^2 - 9 = \pm\sqrt{256} = \pm 16$ ***[1 mark]***
$x^2 = 9 \pm 16 = 25, -7$ ***[1 mark]***
$x = \sqrt{25} = \pm 5$ ***[1 mark]***
You can ignore $x^2 = -7$, as this has no solutions in $x \in \mathbb{R}$.

Pages 18-22: Coordinate Geometry

1 a) To find the coordinates of A, solve the equations of the lines simultaneously:
l_1: $x - y + 1 = 0$
l_2: $2x + y - 8 = 0$
Add the equations to get rid of y:
$3x - 7 = 0$ ***[1 mark]*** $\Rightarrow x = \frac{7}{3}$ ***[1 mark]***
Now put $x = \frac{7}{3}$ back into l_1 to find y:
$\frac{7}{3} - y + 1 = 0 \Rightarrow y = \frac{7}{3} + 1 = \frac{10}{3}$
So A is $\left(\frac{7}{3}, \frac{10}{3}\right)$ ***[1 mark]***

b) There's a lot of information here, so draw a quick sketch to make things a bit clearer:

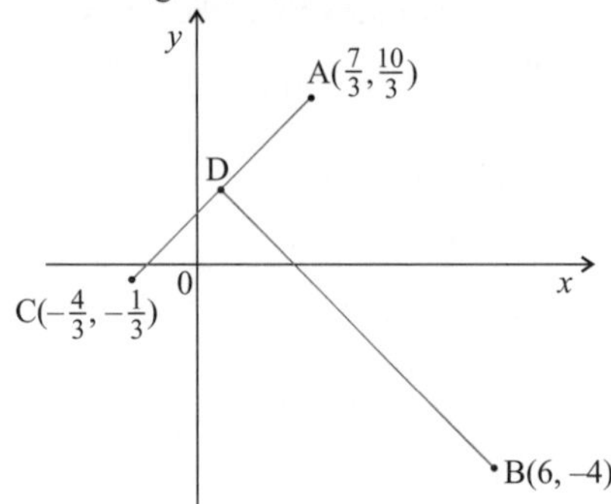

To find the equation of the line through B and D, you need its gradient. But before you can find the gradient, you need to find the coordinates of point D — the midpoint of AC. To find the midpoint of two points, find the average of the x-values and the average of the y-values:

$D = \left(\dfrac{x_A + x_C}{2}, \dfrac{y_A + y_C}{2}\right) = \left(\dfrac{\frac{7}{3} + \frac{-4}{3}}{2}, \dfrac{\frac{10}{3} + \frac{-1}{3}}{2}\right)$ ***[1 mark]***

$D = \left(\frac{1}{2}, \frac{3}{2}\right)$ ***[1 mark]***

To find the gradient (m) of the line through B and D, use this rule: $m_{BD} = \dfrac{y_D - y_B}{x_D - x_B}$

$m = \dfrac{\frac{3}{2} - -4}{\frac{1}{2} - 6} = \dfrac{\frac{3}{2} + \frac{8}{2}}{\frac{1}{2} - \frac{12}{2}} = \dfrac{3+8}{1-12} = -1$ ***[1 mark]***

Now you can find the equation of the line. Input the known values of x and y at B(6, −4) and the gradient (−1) into $y - y_1 = m(x - x_1)$, which gives:
$y - (-4) = -1(x - 6)$ ***[1 mark]***
$\Rightarrow y + 4 = -x + 6 \Rightarrow x + y - 2 = 0$ ***[1 mark]***
You could also have used the point D you found earlier in the question in the formula $y - y_1 = m(x - x_1)$.

c) Look at the sketch in part b). To prove triangle ABD is a right-angled triangle, you need to prove that lines AD and BD are perpendicular — in other words, prove the product of their gradients equals −1.
You already know the gradient of BD = −1.
Use the same rule to find the gradient of AD:

$$m_{AD} = \frac{y_D - y_A}{x_D - x_A}$$

$$m = \frac{\frac{3}{2} - \frac{10}{3}}{\frac{1}{2} - \frac{7}{3}} = \frac{\frac{9}{6} - \frac{20}{6}}{\frac{3}{6} - \frac{14}{6}} = \frac{9-20}{3-14} = 1 \text{ [1 mark]}$$

$m_{BD} \times m_{AD} = -1 \times 1 = -1$ ***[1 mark]***, so triangle ABD is a right-angled triangle ***[1 mark]***.

2 a) The gradient of the line through B and C equals the coefficient of x when the equation of the line is in the form $y = mx + c$.
$-3x + 5y = 16 \Rightarrow 5y = 3x + 16 \Rightarrow y = \frac{3}{5}x + \frac{16}{5}$,
so gradient $= \frac{3}{5}$.
AB and BC are perpendicular, so the gradient of the line through A and B equals $-1 \div \frac{3}{5} = -\frac{5}{3}$ ***[1 mark]***.
Two lines are parallel if they have the same gradient.
$5x + 3y - 6 = 0 \Rightarrow 3y = -5x + 6 \Rightarrow y = -\frac{5}{3}x + 2$,
so the gradient is $-\frac{5}{3}$ ***[1 mark]***.
The line with equation $5x + 3y - 6 = 0$ has the same gradient as the line through points A and B, so the lines are parallel ***[1 mark]***.

b) To calculate the area, find the length of one side — say AB. Point B has coordinates $(3, k)$, so you can find k by substituting $x = 3$ and $y = k$ into the equation of the line through B and C:
$-3x + 5y = 16 \Rightarrow (-3 \times 3) + 5k = 16 \Rightarrow 5k = 25 \Rightarrow k = 5$,
so point B = (3, 5) ***[1 mark]***. Now you can input the values of x and y at B(3, 5) and the gradient $(-\frac{5}{3})$ into $y - y_1 = m(x - x_1)$ to find the equation of AB:
$y - 5 = -\frac{5}{3}(x - 3) \Rightarrow y = -\frac{5}{3}x + 10$ ***[1 mark]***
Use this to find the coordinates of point A:
A lies on the y-axis, so $x = 0$.
When $x = 0$, $y = 10$, so A is the point (0, 10) ***[1 mark]***.
Now find the length AB using Pythagoras' theorem:
$(AB)^2 = (10 - 5)^2 + (0 - 3)^2$ ***[1 mark]*** $= 25 + 9 = 34$,
Area of square $= (AB)^2 = 34$ units^2 ***[1 mark]***

3 a) The centre of the circle must be the midpoint of AB, since AB is a diameter. Midpoint of AB is:
$\left(\frac{2+0}{2}, \frac{1+-5}{2}\right) = (1, -2)$ ***[1 mark]***
The radius is the distance from the centre (1, −2) to point A:
radius $= \sqrt{(2-1)^2 + (1-(-2))^2}$ ***[1 mark]*** $= \sqrt{10}$ ***[1 mark]***

b) The general equation for a circle with centre (a, b) and radius r is: $(x - a)^2 + (y - b)^2 = r^2$. So for a circle with centre (1, −2) and radius $\sqrt{10}$, that gives $(x - 1)^2 + (y + 2)^2 = 10$ ***[1 mark]***.
To show that the point (4, −1) lies on the circle, show that it satisfies the equation of the circle: $(4 - 1)^2 + (-1 + 2)^2 = 9 + 1 = 10$, so (4, −1) lies on the circle ***[1 mark]***.

c) Start with your equation from part b) and multiply out to get the form given in the question:
$(x - 1)^2 + (y + 2)^2 = 10$
$x^2 - 2x + 1 + y^2 + 4y + 4 = 10$
$x^2 + y^2 - 2x + 4y - 5 = 0$
[2 marks available — 1 mark for multiplying out the equation from part b, 1 mark for correct rearrangement to give the answer in the required form]

d) The radius at A has the same gradient as the diameter AB, so gradient of radius $= \frac{1 - -5}{2 - 0} = 3$ ***[1 mark]***.
The tangent at point A is perpendicular to the radius at point A, so the tangent has gradient $-1 \div 3 = -\frac{1}{3}$ ***[1 mark]***.
Put the gradient $-\frac{1}{3}$ and point A(2, 1) into the formula for the equation of a straight line and rearrange:
$y - y_1 = m(x - x_1) \Rightarrow y - 1 = -\frac{1}{3}(x - 2)$
$\Rightarrow y - 1 = -\frac{1}{3}x + \frac{2}{3} \Rightarrow y = -\frac{1}{3}x + \frac{5}{3}$ ***[1 mark]***

4 a) The line through the centre P bisects the chord, and so is perpendicular to the chord AB at the midpoint M.
Gradient of AB = Gradient of AM $= \frac{(7-10)}{(11-9)} = -\frac{3}{2}$.
Gradient of PM $= -1 \div -\frac{3}{2} = \frac{2}{3}$
Gradient of PM $= \frac{(7-3)}{(11-p)} = \frac{2}{3}$
$\Rightarrow 3(7 - 3) = 2(11 - p) \Rightarrow 12 = 22 - 2p \Rightarrow p = 5$.
[5 marks available — 1 mark for identifying that PM and AB are perpendicular, 1 mark for correct gradient of AB (or AM), 1 mark for correct gradient of PM, 1 mark for substitution of the y-coordinate of P into the equation for the gradient or equation of the line PM, and 1 mark for correct rearrangement to give the answer in the required form]

b) The equation of a circle is $(x - a)^2 + (y - b)^2 = r^2$.
The centre of the circle is P(5, 3), so $a = 5$ and $b = 3$ ***[1 mark]***.
r^2 is the square of the radius. The radius equals the length of AP, so you can find r^2 using Pythagoras' theorem:
$r^2 = (AP)^2 = (9 - 5)^2 + (10 - 3)^2$ ***[1 mark]*** $= 65$
So the equation of the circle is:
$(x - 5)^2 + (y - 3)^2 = 65$ ***[1 mark]***

5 a) A is on the y-axis, so the x-coordinate is 0.
Just put $x = 0$ into the equation and solve:
$0^2 - (6 \times 0) + y^2 - 4y = 0$ ***[1 mark]***
$\Rightarrow y^2 - 4y = 0 \Rightarrow y(y - 4) = 0 \Rightarrow y = 0$ or $y = 4$
$y = 0$ is the origin, so A is at (0, 4) ***[1 mark]***

b) Complete the square for the terms involving x and y separately:
Completing the square for $x^2 - 6x$ means you have to start with $(x - 3)^2$, but $(x - 3)^2 = x^2 - 6x + 9$, so you need to subtract 9:
$(x - 3)^2 - 9$ ***[1 mark]***
Now the same for $y^2 - 4y$: $(y - 2)^2 = y^2 - 4y + 4$,
so subtract 4 which gives: $(y - 2)^2 - 4$ ***[1 mark]***
Put these new expressions back into the original equation:
$(x - 3)^2 - 9 + (y - 2)^2 - 4 = 0$
$\Rightarrow (x - 3)^2 + (y - 2)^2 = 13$ ***[1 mark]***

c) In the general equation for a circle $(x - a)^2 + (y - b)^2 = r^2$, the centre is (a, b) and the radius is r.
So for the equation in part b), $a = 3$, $b = 2$, $r = \sqrt{13}$.
Hence, the centre is (3, 2) ***[1 mark]***
and the radius is $\sqrt{13}$ ***[1 mark]***

d) The tangent at point A is perpendicular to the radius at A.
The radius between A(0, 4) and the centre (3, 2) has gradient:
$\frac{y_2 - y_1}{x_2 - x_1} = \frac{2-4}{3-0} = -\frac{2}{3}$ ***[1 mark]***
So the gradient of the tangent at A $= -1 \div -\frac{2}{3} = \frac{3}{2}$ ***[1 mark]***
Put $m = \frac{3}{2}$ and A = (0, 4) into $y - y_1 = m(x - x_1)$ to find the equation of the tangent to the circle at point A:
$y - 4 = \frac{3}{2}(x - 0) \Rightarrow y - 4 = \frac{3}{2}x \Rightarrow y = \frac{3}{2}x + 4$ ***[1 mark]***
Point B lies on the line with equation $y = \frac{3}{2}x + 4$.
B also lies on the x-axis, so substitute $y = 0$ into the equation of the line to find the x-coordinate of B:
$0 = \frac{3}{2}x + 4 \Rightarrow x = -\frac{8}{3}$, so B is the point $\left(-\frac{8}{3}, 0\right)$ ***[1 mark]***
Now find AB using Pythagoras' theorem:
$(AB)^2 = \left(0 - -\frac{8}{3}\right)^2 + (4 - 0)^2$ ***[1 mark]*** $= \frac{64}{9} + 16$,
so AB $= \sqrt{\frac{64}{9} + 16} = \sqrt{\frac{208}{9}} = \frac{4\sqrt{13}}{3}$ ***[1 mark]***
If the question asks for an exact answer, leave it in surd form.

6 First, find the centre of the circle. Do this by finding the perpendicular bisectors of any two sides:
Midpoint of $PQ = \left(\frac{3+0}{2}, \frac{1+2}{2}\right) = \left(\frac{3}{2}, \frac{3}{2}\right)$,
gradient of $PQ = \frac{y_P - y_Q}{x_P - x_Q} = \frac{1-2}{3-0} = \frac{-1}{3} = -\frac{1}{3}$
So the perpendicular bisector of PQ has gradient $-1 \div -\frac{1}{3} = 3$ ***[1 mark]*** and goes through $\left(\frac{3}{2}, \frac{3}{2}\right)$, so has equation $y - \frac{3}{2} = 3(x - \frac{3}{2}) \Rightarrow y = 3x - 3$ ***[1 mark]***.
Midpoint of $PR = \left(\frac{3+1}{2}, \frac{1+5}{2}\right) = (2, 3)$,
gradient of $PR = \frac{y_P - y_R}{x_P - x_R} = \frac{1-5}{3-1} = \frac{-4}{2} = -2$
So the perpendicular bisector of PR has gradient $-1 \div -2 = \frac{1}{2}$ ***[1 mark]*** and goes through (2, 3), so has equation $y - 3 = \frac{1}{2}(x-2) \Rightarrow y = \frac{1}{2}x + 2$ ***[1 mark]***.
Find the centre of the circle by setting these equations of the perpendicular bisectors equal to one another and solving:
$3x - 3 = \frac{1}{2}x + 2 \Rightarrow \frac{5}{2}x = 5 \Rightarrow x = 2$, so $y = (3 \times 2) - 3 = 3$
So the centre is (2, 3) ***[1 mark]***.
The distance from the centre to point Q (0, 2) is the radius:
$r = \sqrt{(2-0)^2 + (3-2)^2} = \sqrt{4+1} = \sqrt{5}$, so $r^2 = (\sqrt{5})^2 = 5$ ***[1 mark]***
So the equation of C is $(x-2)^2 + (y-3)^2 = 5$ ***[1 mark]***.
There are lots of different ways to get the right answer here, depending on which sides or points you use in your calculation.

7 a) At A, $y = 4$, so $4\cos\theta = 4 \Rightarrow \cos\theta = 1$
$\Rightarrow \theta = 0$, as $0 \le \theta \le \frac{\pi}{2}$.
At B, $x = 3$, so $3\sin\theta = 3 \Rightarrow \sin\theta = 1$
$\Rightarrow \theta = \frac{\pi}{2}$, as $0 \le \theta \le \frac{\pi}{2}$.
[2 marks available — 1 mark for each value of θ]

b) $y^2 = 16\cos^2\theta$. Use the identity $\cos^2\theta \equiv 1 - \sin^2\theta$:
$y^2 = 16(1 - \sin^2\theta)$ ***[1 mark]***.
Now, $x^2 = 9\sin^2\theta$, so $\frac{x^2}{9} = \sin^2\theta$. ***[1 mark]***
Substitute this into the equation for y^2:
$y^2 = 16(1 - \frac{x^2}{9}) = 16 - \frac{16x^2}{9} = \left(4 + \frac{4x}{3}\right)\left(4 - \frac{4x}{3}\right)$ ***[1 mark]***
The last step is a difference of squares.

8 For the boat to be further west than the tip of the island, this means $x < 12$, so:
$t^2 - 7t + 12 < 12$ ***[1 mark]*** $\Rightarrow t^2 - 7t < 0 \Rightarrow t(t-7) < 0$ ***[1 mark]***
This means that the boat starts level with the tip of the island (at $t = 0$), and is then level again when $t = 7$. So the boat is west of the tip of the island for 7 hours ***[1 mark]***.

9 a) Substitute the given value of θ into the parametric equations:
$\theta = \frac{\pi}{3} \Rightarrow x = 1 - \tan\frac{\pi}{3} = 1 - \sqrt{3}$
$y = \frac{1}{2}\sin\left(\frac{2\pi}{3}\right) = \frac{1}{2}\left(\frac{\sqrt{3}}{2}\right) = \frac{\sqrt{3}}{4}$
So $P = \left(1 - \sqrt{3}, \frac{\sqrt{3}}{4}\right)$
[2 marks available — 1 mark for substituting θ = $\frac{\pi}{3}$ into the parametric equations, 1 mark for both coordinates of P correct]

b) Use $y = -\frac{1}{2}$ to find the value of θ:
$-\frac{1}{2} = \frac{1}{2}\sin 2\theta \Rightarrow \sin 2\theta = -1$
$\Rightarrow 2\theta = -\frac{\pi}{2} \Rightarrow \theta = -\frac{\pi}{4}$
[2 marks available — 1 mark for substituting given x- or y-value into the correct parametric equation, 1 mark for finding the correct value of θ]
You can also find θ using the parametric equation for x, with x = 2.

c) $x = 1 - \tan\theta \Rightarrow \tan\theta = 1 - x$
$y = \frac{1}{2}\sin 2\theta = \frac{1}{2}\left(\frac{2\tan\theta}{1+\tan^2\theta}\right)$
$= \frac{\tan\theta}{1+\tan^2\theta} = \frac{(1-x)}{1+(1-x)^2} = \frac{1-x}{1+1-2x+x^2}$
$= \frac{1-x}{x^2-2x+2}$
[3 marks available — 1 mark for using the given identity to rearrange one of the parametric equations, 1 mark for eliminating θ from the parametric equation for y, 1 mark for correctly expanding to obtain the Cartesian equation given in the question]

10 a) C crosses the y-axis when $x = 0$,
so when $4t - 2 = 0 \Rightarrow 4t = 2 \Rightarrow t = \frac{1}{2}$ ***[1 mark]***
Substitute this into the equation for y:
$y = \left(\frac{1}{2}\right)^3 + \frac{1}{2} = \frac{5}{8}$, so the coordinates are $\left(0, \frac{5}{8}\right)$ ***[1 mark]***

b) Substitute $x = 4t - 2$ into $y = \frac{1}{2}x + 1$:
$y = \frac{1}{2}(4t-2) + 1 = 2t$. Now solve for t when $y = t^3 + t$:
$2t = t^3 + t \Rightarrow t^3 - t = 0 \Rightarrow t(t^2 - 1) = 0 \Rightarrow t(t-1)(t+1) = 0$.
So $t = 0$, 1 and -1.
When $t = 0$, $x = -2$ and $y = 0$ so the point has coordinates $(-2, 0)$
When $t = 1$, $x = 2$ and $y = 2$ so the point has coordinates (2, 2)
When $t = -1$, $x = -6$ and $y = -2$ so the point has coordinates $(-6, -2)$
[4 marks available — 1 mark for a correct method to find t at points of intersection, 1 mark for all values of t correct, 2 marks for all coordinates correct, otherwise 1 mark for two coordinates correct]

c) $x = 4t - 2 \Rightarrow t = \frac{x+2}{4}$ ***[1 mark]***
Substitute this into the equation for y:
$y = \left(\frac{x+2}{4}\right)^3 + \frac{x+2}{4} = \frac{x^3 + 6x^2 + 12x + 8}{64} + \frac{x+2}{4}$ ***[1 mark]***
$= \frac{x^3 + 6x^2 + 28x + 40}{64} = \frac{1}{64}x^3 + \frac{3}{32}x^2 + \frac{7}{16}x + \frac{5}{8}$ ***[1 mark]***
(so $a = \frac{1}{64}$, $b = \frac{3}{32}$, $c = \frac{7}{16}$ and $d = \frac{5}{8}$)

Pages 23-25: Sequences and Series 1

1 a) In an arithmetic series, the nth term is defined by the formula $a + (n-1)d$. The 12th term is 79, so the equation is $79 = a + 11d$, and the 16th term is 103, so the other equation is $103 = a + 15d$ ***[1 mark for both equations]***. Solving these simultaneously (by taking the first equation away from the second) gives $24 = 4d$, so $d = 6$ ***[1 mark]***.
Putting this value of d into the first equation gives $79 = a + (11 \times 6)$, so $a = 13$. ***[1 mark]***

b) $S_n = \frac{n}{2}[2a + (n-1)d]$
Putting in the values of a and d from above, and $n = 15$:
$S_{15} = \frac{15}{2}[26 + 14 \times 6]$ ***[1 mark]*** $= 825$ ***[1 mark]***
The formula for S_n is given on the formula pages.

2 a) Work out the terms one by one:
$x_1 = k$
$x_2 = 3k - 4$
$x_3 = 3(3k-4) - 4 = 9k - 16$ ***[1 mark]***
$x_4 = 3(9k-16) - 4 = 27k - 52$ ***[1 mark]***

b) Using the terms above with $k = 1$, the sequence is:
1, –1, –7, –25, ... ***[1 mark]***
As the sequence involves multiplying the previous number by three and then taking away a number, once it is negative, it will only decrease (it'll become a larger negative number), therefore the sequence is decreasing ***[1 mark]***.

3 a) $h_2 = 2h_1 + 2 = 2 \times 5 + 2 = 12$
$h_3 = 2h_2 + 2 = 2 \times 12 + 2 = 26$
$h_4 = 2h_3 + 2 = 2 \times 26 + 2 = 54$
[2 marks available — 1 mark for a correct method, 1 mark for all three correct answers]

b) $\sum_{r=3}^{6} h_r = h_3 + h_4 + h_5 + h_6$
$h_5 = 2h_4 + 2 = 2 \times 54 + 2 = 110$ and
$h_6 = 2h_5 + 2 = 2 \times 110 + 2 = 222$ ***[1 mark for both correct]***
So $\sum_{r=3}^{6} h_r = 26 + 54 + 110 + 222$ ***[1 mark]*** $= 412$ ***[1 mark]***

4 a) First put the two known terms into the formula for the nth term of a geometric series, $u_n = ar^{n-1}$:
$u_3 = ar^2 = \frac{5}{2}$ and $u_6 = ar^5 = \frac{5}{16}$ ***[1 mark for both]***
Divide the expression for u_6 by the expression for u_3 to get an expression just containing r and solve it:
$\frac{ar^5}{ar^2} = \frac{5}{16} \div \frac{5}{2} \Rightarrow r^3 = \frac{5}{16} \times \frac{2}{5} = \frac{10}{80}$
$r^3 = \frac{1}{8} \Rightarrow r = \sqrt[3]{\frac{1}{8}} \Rightarrow r = \frac{1}{2}$ ***[1 mark]***
Put this value back into the expression for u_3 to find a:
$a\left(\frac{1}{2}\right)^2 = \frac{5}{2} \Rightarrow \frac{a}{4} = \frac{5}{2} \Rightarrow a = 10$ ***[1 mark]***
The nth term is $u_n = ar^{n-1}$, where $r = \frac{1}{2}$ and $a = 10$
$u_n = 10 \times \left(\frac{1}{2}\right)^{n-1} = 10 \times \frac{1}{2^{n-1}} = \frac{10}{2^{n-1}}$ ***[1 mark]***

b) $S_n = \frac{a(1-r^n)}{1-r}$,
$S_{10} = \frac{10\left(1-\left(\frac{1}{2}\right)^{10}\right)}{1-\left(\frac{1}{2}\right)}$ ***[1 mark]*** $= 10 \times 2 \times \left(1 - \frac{1}{2^{10}}\right)$
$= 20\left(1 - \frac{1}{1024}\right) = \frac{5115}{256}$ ***[1 mark]***

c) Substitute $a = 10$ and $r = \frac{1}{2}$ into the sum to infinity formula:
$S_\infty = \frac{a}{1-r} = \frac{10}{1-\left(\frac{1}{2}\right)} = 10 \div \frac{1}{2} = 10 \times 2 = 20$
[2 marks available — 1 mark for a correct method and 1 mark for showing the sum to infinity is 20]

5 a) The series is defined by $u_{n+1} = 5 \times 1.7^n$ so $r = 1.7$.
r is greater than 1, so the sequence is divergent, which means the sum to infinity cannot be found ***[1 mark]***.

b) $u_3 = 5 \times 1.7^2 = 14.45$ ***[1 mark]***
$u_8 = 5 \times 1.7^7 = 205.17$ (2 d.p.) ***[1 mark]***

6 a) $S_\infty = \frac{a}{1-r} = \frac{20}{1-\frac{3}{4}} = \frac{20}{\frac{1}{4}} = 80$
[2 marks available — 1 mark for substituting into the correct formula, 1 mark for correct answer]

b) $u_{15} = ar^{14} = 20 \times \left(\frac{3}{4}\right)^{14} = 0.356$ (to 3 s.f.)
[2 marks available — 1 mark for substituting into the correct formula, 1 mark for correct answer]

c) Use the formula for the sum of a geometric series to write an expression for S_n:
$S_n = \frac{a(1-r^n)}{1-r} = \frac{20\left(1-\left(\frac{3}{4}\right)^n\right)}{1-\frac{3}{4}}$ ***[1 mark]***
so $\frac{20\left(1-\left(\frac{3}{4}\right)^n\right)}{1-\frac{3}{4}} > 79.76$
Now rearrange and use logs to get n on its own:
$\frac{20\left(1-\left(\frac{3}{4}\right)^n\right)}{1-\frac{3}{4}} > 79.76 \Rightarrow 20\left(1-\left(\frac{3}{4}\right)^n\right) > 19.94$
$\Rightarrow 1-\left(\frac{3}{4}\right)^n > 0.997 \Rightarrow 0.003 > 0.75^n$ ***[1 mark]***
$\Rightarrow \log 0.003 > n \log 0.75$ ***[1 mark]***
$\Rightarrow \frac{\log 0.003}{\log 0.75} < n$ ***[1 mark]***
Remember — if x < 1, then log x has a negative value and dividing by a negative means flipping the inequality.
$\frac{\log 0.003}{\log 0.75} = 20.1929....$
so $n > 20.1929....$
But n must be an integer, so $n = 21$ ***[1 mark]***

7 a) Use $u_n = ar^{n-1}$ with $a = 1$ and $r = 1.5$:
$u_5 = 1 \times (1.5)^4$ ***[1 mark]***
$= 5.06$ km (to the nearest 10 m) ***[1 mark]***

b) Use $u_n = ar^{n-1}$ with $a = 2$ and $r = 1.2$:
$u_9 = 2 \times (1.2)^8 = 8.60$ km (3 s.f.) ***[1 mark]***
$u_{10} = 2 \times (1.2)^9 = 10.3$ km (3 s.f.) ***[1 mark]***
$u_9 < 10$ km and $u_{10} > 10$ km, so day 10 is the first day that Chris runs further than 10 km. ***[1 mark]***
You could've used logs to solve $2(1.2)^{n-1} > 10$ here instead.

c) Alex: $3 \times 10 = 30$ km ***[1 mark]***
Use the formula for the sum of first n terms: $S_n = \frac{a(1-r^n)}{1-r}$
Chris: $\frac{2(1-1.2^{10})}{1-1.2} = 51.917...$ km ***[1 mark]***
Heather: $\frac{1(1-1.5^{10})}{1-1.5} = 113.330...$ km ***[1 mark]***
Total raised $= 30 + 51.917... + 113.330...$
$=$ £195.25 (to the nearest penny) ***[1 mark]***

8 a) $S_\infty = \frac{a}{1-r} = -9 \Rightarrow a = -9(1-r)$
and $ar = -2 \Rightarrow a = \frac{-2}{r}$
$\Rightarrow \frac{-2}{r} = -9(1-r)$
$\Rightarrow -2 = -9r + 9r^2 \Rightarrow 9r^2 - 9r + 2 = 0$
[3 marks available — 1 mark for finding two expressions in a and r, 1 mark for setting these expressions equal to each other, 1 mark for rearranging to give answer in required form]

b) $9r^2 - 9r + 2 = 0 \Rightarrow (3r-1)(3r-2) = 0$ ***[1 mark]***
$\Rightarrow r = \frac{1}{3}$ or $r = \frac{2}{3}$ ***[1 mark for both]***

c) $ar = -2 \Rightarrow a = \frac{-2}{r}$
$r = \frac{1}{3} \Rightarrow a = \frac{-2}{\frac{1}{3}} = -6$ ***[1 mark]***
$r = \frac{2}{3} \Rightarrow a = \frac{-2}{\frac{2}{3}} = -3$ ***[1 mark]***

Pages 26-29: Sequences and Series 2

1 a) Expand $(1 + ax)^{10}$ using the binomial expansion formula:
$(1 + ax)^{10} = 1 + \frac{10}{1}(ax) + \frac{10 \times 9}{1 \times 2}(ax)^2 + \frac{10 \times 9 \times 8}{1 \times 2 \times 3}(ax)^3 + ...$
$= 1 + 10ax + 45a^2x^2 + 120a^3x^3$
[2 marks available — 1 mark for substituting into the formula correctly, 1 mark for correct answer]

b) First take out a factor of 2 to get it in the form $(1 + ax)^n$:
$(2 + 3x)^5 = \left[2\left(1 + \frac{3}{2}x\right)\right]^5 = 2^5\left(1 + \frac{3}{2}x\right)^5 = 32\left(1 + \frac{3}{2}x\right)^5$
[1 mark]
Now expand: $32\left[1 + \frac{5}{1}\left(\frac{3}{2}x\right) + \frac{5 \times 4}{1 \times 2}\left(\frac{3}{2}x\right)^2 + ...\right]$
You only need the x^2 term, so simplify that one:
$32 \times \frac{5 \times 4}{1 \times 2} \times \left(\frac{3}{2}\right)^2 \times x^2$ ***[1 mark]*** $= 720x^2$
So the coefficient of x^2 is 720 ***[1 mark]***

c) Look back to the x^2 term from part a) — it's $45a^2x^2$.
This is equal to $720x^2$ so just rearrange the equation to find a:
$45a^2 = 720 \Rightarrow a^2 = 16 \Rightarrow a = \pm 4$
From part a), $a > 0$, so $a = 4$ ***[1 mark]***

2 a) c represents the coefficient of x^3, so find an expression for the coefficient of x^3 using the binomial expansion formula:
$(j + kx)^6 = j^6\left(1 + \frac{k}{j}x\right)^6$
Coefficient of $x^3 = j^6 \times \frac{6 \times 5 \times 4}{1 \times 2 \times 3} \times \left(\frac{k}{j}\right)^3$ ***[1 mark]***
so $j^6 \times \frac{6 \times 5 \times 4}{1 \times 2 \times 3} \times \left(\frac{k}{j}\right)^3 = 20\,000$ ***[1 mark]***
$j^6 \times 20 \times \left(\frac{1}{j^3}\right) \times k^3 = 20\,000$
$j^6 \times j^{-3} \times k^3 = 1000 \Rightarrow j^3 \times k^3 = 1000 \Rightarrow (jk)^3 = 1000$,
so $jk = \sqrt[3]{1000} = 10$ ***[1 mark for correct rearrangement]***

b) Write an expression for the coefficient of x and then solve simultaneously with the equation $jk = 10$.

coefficient of x: $j^6 \times \frac{6}{1} \times \frac{k}{j} = 37\,500$ ***[1 mark]***

$j^6 \times j^{-1} \times k \times 6 = 37\,500 \Rightarrow kj^5 = 6250$

From a), $jk = 10$, so $k = \frac{10}{j}$

$kj^5 = \frac{10}{j} \times j^5 = 6250$ ***[1 mark for using jk = 10]***

$\Rightarrow 10 \times j^{-1} \times j^5 = 6250 \Rightarrow j^4 = 625 \Rightarrow j = \pm 5$

But j and k are positive so $j = 5$ ***[1 mark]***

Now input $j = 5$ into $jk = 10$ to find: $k = 2$ ***[1 mark]***

c) Coefficient of x^2: $b = 5^6 \times \frac{6 \times 5}{1 \times 2} \times \left(\frac{2}{5}\right)^2 = 37\,500$

[2 marks available — 1 mark for formula, 1 mark for correct answer]

3 a) Using the binomial expansion,

$$(1-x)^{-\frac{1}{2}} \approx 1 + \left(-\frac{1}{2}\right)(-x) + \frac{(-\frac{1}{2})\times(-\frac{3}{2})}{1\times 2}(-x)^2 + \frac{(-\frac{1}{2})\times(-\frac{3}{2})\times(-\frac{5}{2})}{1\times 2\times 3}(-x)^3 \quad \textbf{\textit{[1 mark]}}$$

$$= 1 + \frac{x}{2} + \frac{3}{8}x^2 + \frac{5}{16}x^3.$$

So $A = \frac{1}{2}$, $B = \frac{3}{8}$, $C = \frac{5}{16}$ ***[1 mark]***

b) (i) $(25 - 4x)^{-\frac{1}{2}}$

$$= (25)^{-\frac{1}{2}}\left(1 - \frac{4}{25}x\right)^{-\frac{1}{2}} = \frac{1}{5}\left(1 - \frac{4}{25}x\right)^{-\frac{1}{2}} \quad \textbf{\textit{[1 mark]}}$$

$$= \frac{1}{5}\left(1 + \frac{1}{2}\left(\frac{4}{25}x\right) + \frac{3}{8}\left(\frac{4}{25}x\right)^2 + \frac{5}{16}\left(\frac{4}{25}x\right)^3\right) \quad \textbf{\textit{[1 mark]}}$$

$$= \frac{1}{5}\left(1 + \frac{1}{2}\left(\frac{4}{25}x\right) + \frac{3}{8}\left(\frac{16}{625}x^2\right) + \frac{5}{16}\left(\frac{64}{15\,625}x^3\right)\right)$$

$$= \frac{1}{5}\left(1 + \frac{2}{25}x + \frac{6}{625}x^2 + \frac{4}{3125}x^3\right) \quad \textbf{\textit{[1 mark]}}$$

$$= \frac{1}{5} + \frac{2}{125}x + \frac{6}{3125}x^2 + \frac{4}{15\,625}x^3 \quad \textbf{\textit{[1 mark]}}$$

(ii) The expansion is valid for

$$\left|\frac{-4x}{25}\right| < 1 \Rightarrow \frac{|-4\|x|}{25} < 1 \Rightarrow |x| < \frac{25}{4} \quad \textbf{\textit{[1 mark]}}$$

c) (i) $25 - 4x = 20 \Rightarrow x = \frac{5}{4}$ ***[1 mark]***

$$\text{So } \frac{1}{\sqrt{20}} = \left(25 - 4\left(\frac{5}{4}\right)\right)^{-\frac{1}{2}}$$

$$\approx \frac{1}{5} + \frac{2}{125}\left(\frac{5}{4}\right) + \frac{6}{3125}\left(\frac{5}{4}\right)^2 + \frac{4}{15\,625}\left(\frac{5}{4}\right)^3 \quad \textbf{\textit{[1 mark]}}$$

$$= \frac{1}{5} + \frac{2}{125}\left(\frac{5}{4}\right) + \frac{6}{3125}\left(\frac{25}{16}\right) + \frac{4}{15\,625}\left(\frac{125}{64}\right)$$

$$= \frac{1}{5} + \frac{1}{50} + \frac{3}{1000} + \frac{1}{2000}$$

$$= \frac{447}{2000}$$ ***[1 mark for correct simplification]***

(ii) Percentage error

$$= \left|\frac{\text{real value} - \text{estimate}}{\text{real value}}\right| \times 100$$

$$= \left|\frac{\frac{1}{\sqrt{20}} - \frac{447}{2000}}{\frac{1}{\sqrt{20}}}\right| \times 100 \quad \textbf{\textit{[1 mark]}}$$

$= 0.0004776... = 0.0478\%$ (3 s.f.) ***[1 mark]***

4 a) $(27 + 4x)^{\frac{1}{3}} = 27^{\frac{1}{3}}\left(1 + \frac{4}{27}x\right)^{\frac{1}{3}} = 3\left(1 + \frac{4}{27}x\right)^{\frac{1}{3}}$ ***[1 mark]***

$$\approx 3\left[1 + \left(\frac{1}{3}\right)\left(\frac{4}{27}x\right) + \frac{\frac{1}{3}\times -\frac{2}{3}}{1\times 2}\left(\frac{4}{27}x\right)^2\right] \quad \textbf{\textit{[1 mark]}}$$

$$= 3\left[1 + \left(\frac{1}{3}\right)\left(\frac{4}{27}x\right) + \left(-\frac{1}{9}\right)\left(\frac{16}{729}x^2\right)\right]$$

$$= 3\left(1 + \frac{4}{81}x - \frac{16}{6561}x^2\right) \quad \textbf{\textit{[1 mark]}}$$

$$= 3 + \frac{4}{27}x - \frac{16}{2187}x^2 \quad \textbf{\textit{[1 mark]}}$$

b) $27 + 4x = 26.2 \Rightarrow x = -0.2$

This lies within $|x| < \frac{16}{3}$, so it's a valid approximation, and x is small, so higher powers can be ignored.

$$\sqrt[3]{26.2} = (26.2)^{\frac{1}{3}} \approx 3 + \frac{4}{27}(-0.2) - \frac{16}{2187}(-0.2)^2$$

$= 3 - 0.0296296... - 0.0002926...$

$= 2.9700777... = 2.970078$ (6 d.p.)

[2 marks available — 1 mark for substituting x = –0.2 into the expansion from part a), 1 mark for correct answer]

5 a) (i) $\sqrt{\frac{1+3x}{1-5x}} = \frac{\sqrt{1+3x}}{\sqrt{1-5x}}$

$= (1 + 3x)^{\frac{1}{2}}(1 - 5x)^{-\frac{1}{2}}$ ***[1 mark]***

$$(1+3x)^{\frac{1}{2}} = 1 + \frac{1}{2}(3x) + \frac{\frac{1}{2}\times -\frac{1}{2}}{1\times 2}(3x)^2 + \ldots$$

$$\approx 1 + \frac{3}{2}x - \frac{9}{8}x^2 \quad \textbf{\textit{[1 mark]}}$$

$$(1-5x)^{-\frac{1}{2}} = 1 + \left(-\frac{1}{2}\right)(-5x) + \frac{-\frac{1}{2}\times -\frac{3}{2}}{1\times 2}(-5x)^2 + \ldots$$

$$\approx 1 + \frac{5}{2}x + \frac{75}{8}x^2 \quad \textbf{\textit{[1 mark]}}$$

$$\sqrt{\frac{1+3x}{1-5x}} \approx \left(1 + \frac{3}{2}x - \frac{9}{8}x^2\right)\left(1 + \frac{5}{2}x + \frac{75}{8}x^2\right) \quad \textbf{\textit{[1 mark]}}$$

$$\approx 1 + \frac{5}{2}x + \frac{75}{8}x^2 + \frac{3}{2}x + \frac{15}{4}x^2 - \frac{9}{8}x^2$$

(ignoring any terms in x^3 or above)

$= 1 + 4x + 12x^2$ ***[1 mark for correct simplification]***

(ii) Expansion of $(1 + 3x)^{\frac{1}{2}}$ is valid for: $|3x| < 1 \Rightarrow |x| < \frac{1}{3}$.

Expansion of $(1 - 5x)^{-\frac{1}{2}}$ is valid for:

$|-5x| < 1 \Rightarrow |-5||x| < 1 \Rightarrow |x| < \frac{1}{5}$.

The combined expansion is valid for the narrower of these two ranges, so the expansion of $\sqrt{\frac{1+3x}{1-5x}}$ is valid for: $|x| < \frac{1}{5}$.

[2 marks available — 1 mark for identifying the valid range of the expansion as being the narrower of the two valid ranges shown, 1 mark for correct answer]

b) $x = \frac{1}{15} \Rightarrow \sqrt{\frac{1+3x}{1-5x}} = \sqrt{\frac{1+\frac{3}{15}}{1-\frac{5}{15}}} = \sqrt{\frac{\left(\frac{18}{15}\right)}{\left(\frac{10}{15}\right)}} = \sqrt{\frac{18}{10}} = \sqrt{1.8}$

$$\sqrt{1.8} \approx 1 + 4\left(\frac{1}{15}\right) + 12\left(\frac{1}{15}\right)^2 = 1 + \frac{4}{15} + \frac{4}{75} = \frac{33}{25}$$

[2 marks available — 1 mark for substituting $x = \frac{1}{15}$ into the expansion from part a), 1 mark for correct simplification]

6 a) $2 - 18x \equiv A(1 - 2x)^2 + B(5 + 4x)(1 - 2x) + C(5 + 4x)$ ***[1 mark]***

Let $x = \frac{1}{2}$, then:

$2 - 9 = 7C \Rightarrow -7 = 7C \Rightarrow C = -1$ ***[1 mark]***

Let $x = -\frac{5}{4}$, then:

$2 - -\frac{45}{2} = \frac{49}{4}A \Rightarrow \frac{49}{2} = \frac{49}{4}A \Rightarrow A = 2$ ***[1 mark]***

Equating the coefficients of the x^2 terms:

$0 = 4A - 8B = 8 - 8B \Rightarrow 8B = 8 \Rightarrow B = 1$ ***[1 mark]***

b) $f(x) = \frac{2}{(5+4x)} + \frac{1}{(1-2x)} - \frac{1}{(1-2x)^2}$

$= 2(5 + 4x)^{-1} + (1 - 2x)^{-1} - (1 - 2x)^{-2}$

Expand each bracket separately:

$$(5+4x)^{-1} = 5^{-1}\left(1 + \frac{4}{5}x\right)^{-1} = \frac{1}{5}\left(1 + \frac{4}{5}x\right)^{-1}$$

$$= \frac{1}{5}\left(1 + (-1)\left(\frac{4}{5}x\right) + \frac{-1\times -2}{1\times 2}\left(\frac{4}{5}x\right)^2 + \ldots\right)$$

$$= \frac{1}{5}\left(1 - \frac{4}{5}x + \frac{16}{25}x^2 + \ldots\right) = \frac{1}{5} - \frac{4}{25}x + \frac{16}{125}x^2 + \ldots$$

$$(1-2x)^{-1} = 1 + (-1)(-2x) + \frac{-1\times -2}{1\times 2}(-2x)^2 + \ldots$$

$= 1 + 2x + 4x^2 + \ldots$

$$(1-2x)^{-2} = 1 + (-2)(-2x) + \frac{-2\times -3}{1\times 2}(-2x)^2 + \ldots$$

$= 1 + 4x + 12x^2 + \ldots$

Putting it all together gives (ignoring any terms in x^3 or above):

$$f(x) \approx 2\left(\frac{1}{5} - \frac{4}{25}x + \frac{16}{125}x^2\right) + (1 + 2x + 4x^2) - (1 + 4x + 12x^2)$$
$$= \frac{2}{5} - \frac{8}{25}x + \frac{32}{125}x^2 + 1 + 2x + 4x^2 - 1 - 4x - 12x^2$$
$$= \frac{2}{5} - \frac{58}{25}x - \frac{968}{125}x^2$$

[7 marks available — 1 mark for rewriting f(x) in the form $A(5 + 4x)^{-1} + B(1 - 2x)^{-1} + C(1 - 2x)^{-2}$, 1 mark for taking out a factor of 5 from $(5 + 4x)^{-1}$, 1 mark for correct expansion of $(5 + 4x)^{-1}$, 1 mark for correct expansion of $(1 - 2x)^{-1}$, 1 mark for correct expansion of $(1 - 2x)^{-2}$, 1 mark for correct constant and x-terms in final answer, 1 mark for correct x^2-term in final answer]

c) Expansion of $(5 + 4x)^{-1}$ is valid for

$$\left|\frac{4x}{5}\right| < 1 \Rightarrow \frac{4|x|}{5} < 1 \Rightarrow |x| < \frac{5}{4}$$

Expansions of $(1 - 2x)^{-1}$ and $(1 - 2x)^{-2}$ are valid for

$$\left|\frac{-2x}{1}\right| < 1 \Rightarrow \frac{2|x|}{1} < 1 \Rightarrow |x| < \frac{1}{2}$$ ***[1 mark for both ranges]***

The combined expansion is valid for the narrower of these two ranges. So the expansion of f(x) is valid for $|x| < \frac{1}{2}$, which is the same as $-\frac{1}{2} < x < \frac{1}{2}$ ***[1 mark]***.

Pages 30-36: Trigonometry

1 To find the area of the sector, you need the angle in radians:

$(120° \div 180°) \times \pi = \frac{2\pi}{3}$ radians ***[1 mark]***

Now use the arc length to find r:

Arc length $S = r\theta$, so

$40 = \frac{2\pi}{3} \times r$ ***[1 mark]***

$r = 40 \div \frac{2\pi}{3} = \frac{60}{\pi}$ ***[1 mark]***

Finally, use this value of r to find the area:

$A = \frac{1}{2}r^2\theta = \frac{1}{2} \times (\frac{60}{\pi})^2 \times \frac{2\pi}{3}$ ***[1 mark]***

$= \frac{1200}{\pi} = 381.9718... = 382 \text{ cm}^2$ (to the nearest cm²) ***[1 mark]***

2 a) (i) In the diagram, x is the adjacent side of a right-angled triangle with an angle θ and hypotenuse r, so use the cos formula:

$$\cos\theta = \frac{\text{adjacent}}{\text{hypotenuse}} = \frac{x}{r}, \text{ so } x = r\cos\theta$$ ***[1 mark]***

(ii) As the stage is symmetrical, you know that distance y is the same on both triangles.

$$\sin\theta = \frac{\text{opposite}}{\text{hypotenuse}} = \frac{y}{r}, \text{ so } y = r\sin\theta$$ ***[1 mark]***

b) The total length of the bottom and straight sides is $q + q + 2r$. The top length is $2x$, so using the expression found in a), you can write this as $2r\cos\theta$.

For the curved lengths, the shaded areas are sectors of circles, and the formula for the length of one arc is given by $r\theta$.

Now add them all up to get the total perimeter:

$q + q + 2r + 2r\cos\theta + r\theta + r\theta = 2[q + r(1 + \theta + \cos\theta)]$.

Break the area down into a rectangle, a triangle and two sectors:

Area of rectangle = width × height = $2qr$

Area of triangle = $\frac{1}{2}(2r\cos\theta)(r\sin\theta) = r^2\cos\theta\sin\theta$

Area of one shaded sector = $\frac{1}{2}r^2\theta$

So the total area $A = 2qr + r^2\cos\theta\sin\theta + r^2\theta$
$= 2qr + r^2(\cos\theta\sin\theta + \theta)$.

[4 marks available — 1 mark for all individual lengths correct, 1 mark for all individual areas correct, 1 mark for each correct expression]

You could've used $\frac{1}{2}$ AB sin C to find the area of the triangle, but then you'd need to use one of the expressions for x or y from part a) to get the final answer.

c) Substitute the given values of P and θ into the expression for the perimeter: $P = 2[q + r(1 + \theta + \cos\theta)]$

$$\Rightarrow 40 = 2\left[q + r\left(1 + \frac{\pi}{3} + \cos\frac{\pi}{3}\right)\right]$$

$$\Rightarrow 20 = 20 = q + r\left(\frac{3}{2} + \frac{\pi}{3}\right)$$ ***[1 mark]***

And then into the expression for the area:

$A = 2qr + r^2(\cos\theta\sin\theta + \theta)$

$$\Rightarrow A = 2qr + r^2\left(\cos\frac{\pi}{3}\sin\frac{\pi}{3} + \frac{\pi}{3}\right)$$
$$= 2qr + r^2\left(\frac{\sqrt{3}}{4} + \frac{\pi}{3}\right)$$ ***[1 mark]***

To rearrange this formula for area into the form shown in the question, you need to get rid of q. Rearrange the perimeter formula to get an expression for q in terms of r, then substitute that into the area expression:

$$q = 20 - r\left(\frac{3}{2} + \frac{\pi}{3}\right)$$

$$A = 2qr + r^2\left(\frac{\sqrt{3}}{4} + \frac{\pi}{3}\right)$$
$$= 2r\left[20 - r\left(\frac{3}{2} + \frac{\pi}{3}\right)\right] + r^2\left(\frac{\sqrt{3}}{4} + \frac{\pi}{3}\right)$$ ***[1 mark]***
$$= 40r - 2r^2\left(\frac{3}{2} + \frac{\pi}{3}\right) + r^2\left(\frac{\sqrt{3}}{4} + \frac{\pi}{3}\right)$$
$$= 40r - r^2\left[\left(3 + \frac{2\pi}{3}\right) - \left(\frac{\sqrt{3}}{4} + \frac{\pi}{3}\right)\right]$$
$$= 40 - r^2\left(3 - \frac{\sqrt{3}}{4} + \frac{\pi}{3}\right)$$

So $A = 40r - kr^2$, where $k = 3 - \frac{\sqrt{3}}{4} + \frac{\pi}{3}$, as required ***[1 mark]***

3 The graph of $y = \sin x$ is mapped onto the graph of $y = \sin\frac{x}{2}$ via a stretch parallel to the x-axis of scale factor 2.

The graph should appear as follows:

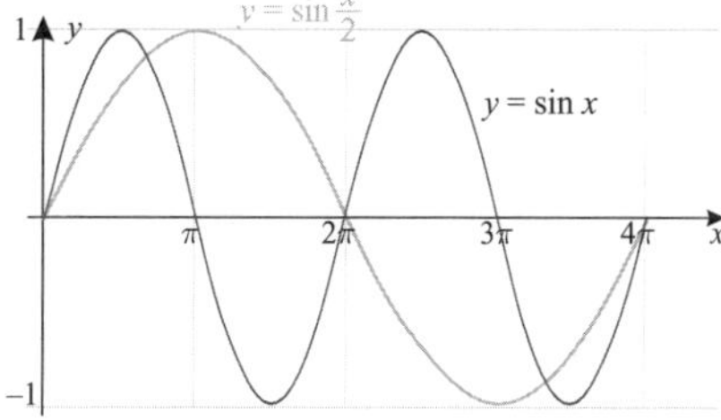

[3 marks available — 1 mark for sin x correct, 1 mark for sin $\frac{x}{2}$ correct, 1 mark for correct axis labelling]

4 a) Use the cosine rule:

e.g. $\cos A = \frac{50^2 + 70^2 - 90^2}{2 \times 50 \times 70}$ ***[1 mark]***

$\cos A = -0.1$ ***[1 mark]***

$A = 95.739...°$ ***[1 mark]***

Now use this value of A to find the area:

Area $= \frac{1}{2} \times 50 \times 70 \times \sin 95.739...°$ ***[1 mark]***

$= 1741.228... = 1741 \text{ m}^2$ (nearest m²) ***[1 mark]***

If you've allocated your values of a, b, c etc. differently, or found a different angle, then the numbers in your working will be different.

b) E.g. the model is unlikely to give an area accurate to the nearest square metre as the given side lengths are most likely rounded, at least to the nearest metre, possibly to the nearest 5 m or 10 m. This means that there is a large range of possible areas. / The sides are unlikely to be perfectly straight, so the model will not be accurate ***[1 mark for a sensible comment]***.

5 $7 - 3\cos x = 9\sin^2 x$, and $\sin^2 x \equiv 1 - \cos^2 x$

$\Rightarrow 7 - 3\cos x = 9(1 - \cos^2 x)$

$\Rightarrow 7 - 3\cos x = 9 - 9\cos^2 x$

$\Rightarrow 9\cos^2 x - 3\cos x - 2 = 0$

Substitute y for $\cos x$ and solve $9y^2 - 3y - 2 = 0$ by factorising:

$(3y - 2)(3y + 1) = 0 \Rightarrow y = \frac{2}{3}$ or $y = -\frac{1}{3}$

So $\cos x = \frac{2}{3}$ or $\cos x = -\frac{1}{3}$

For $\cos x = \frac{2}{3}$, $x = 48.189...° = 48.2°$ (1 d.p.).

For $\cos x = -\frac{1}{3}$, $x = 109.471...° = 109.5°$ (1 d.p).

[5 marks available — 1 mark for correct substitution using trig identity, 1 mark for forming a quadratic in cos x, 1 mark for finding correct values of cos x, 1 mark for each of the 2 correct solutions]

6 a) Substituting $t = 35.26...°$ into both sides of the equation gives:
LHS: $\sin(2 \times 35.26...°) = 0.94$ (2 s.f)
RHS: $\sqrt{2}\cos(2 \times 35.26...°) = 0.47$ (2 s.f.)
$0.94 \neq 0.47$, so Adam's solution is incorrect ***[1 mark]***.

b) Adam has incorrectly divided by 2:
$\tan 2t = \sqrt{2} \not\Rightarrow \tan t = \frac{\sqrt{2}}{2}$ ***[1 mark]***

c) $t = -27.36...°$ is not a solution of the original equation ***[1 mark]***. The error appeared because Bethan has squared the equation and then taken roots ***[1 mark]***.

7 a) Use the trig identity $\tan\theta \equiv \frac{\sin\theta}{\cos\theta}$:
$\tan^2\theta + \frac{\tan\theta}{\cos\theta} = 1 \Rightarrow \frac{\sin^2\theta}{\cos^2\theta} + \frac{\sin\theta}{\cos^2\theta} = 1$ ***[1 mark]***
Put over a common denominator: $\frac{\sin^2\theta + \sin\theta}{\cos^2\theta} = 1$
$\Rightarrow \sin^2\theta + \sin\theta = \cos^2\theta$
Now use the identity $\cos^2\theta \equiv 1 - \sin^2\theta$ to give:
$\sin^2\theta + \sin\theta = 1 - \sin^2\theta$ ***[1 mark]***
$\Rightarrow 2\sin^2\theta + \sin\theta - 1 = 0$ ***[1 mark for rearrangement]***.

b) Factorising the quadratic from a) gives:
$(2\sin\theta - 1)(\sin\theta + 1) = 0$ ***[1 mark]***
$\Rightarrow \sin\theta = \frac{1}{2}$ or $\sin\theta = -1$ ***[1 mark]***
$\sin\theta = \frac{1}{2} \Rightarrow \theta = \frac{\pi}{6}$ and $\theta = (\pi - \frac{\pi}{6}) = \frac{5\pi}{6}$ ***[1 mark for both]***
$\sin\theta = -1 \Rightarrow \theta = \frac{3\pi}{2}$ ***[1 mark]***
So the solutions are $\theta = \frac{\pi}{6}, \frac{5\pi}{6}$ and $\frac{3\pi}{2}$

Don't forget that θ has to be between 0 and 2π — that last value of θ will come up as $-1.5707...$ (which is $-\frac{\pi}{2}$) on your calculator, so you have to work out $\theta = 2\pi - \frac{\pi}{2} = \frac{3\pi}{2}$.

8 a) The start and end points of the cos curve (with restricted domain) are (0, 1) and $(\pi, -1)$, so the coordinates of the start point of arccos (point A) are $(-1, \pi)$ ***[1 mark]*** and the coordinates of the end point (point B) are (1, 0) ***[1 mark]***.

b) $y = \arccos x \Rightarrow y = \cos^{-1}x \Rightarrow x = \cos y$ ***[1 mark]***

c) $\arccos x = 2 \Rightarrow x = \cos 2$ ***[1 mark]*** $\Rightarrow x = -0.416$ ***[1 mark]***

9 a) $\text{cosec}\,\theta = \frac{5}{3} \Rightarrow \sin\theta = \frac{3}{5}$. Solving for θ gives $\theta = 0.64350...$,
$\theta = \pi - 0.64350... = 2.49809...$
So $\theta = 0.644, 2.50$ (both to 3 s.f.)
[1 mark for each correct answer].

Sketch the graph of y = sin x or use a CAST diagram to help you find the second solution.

b) (i) The identity $\text{cosec}^2\theta \equiv 1 + \cot^2\theta$ rearranges to give $\text{cosec}^2\theta - 1 \equiv \cot^2\theta$. Putting this into the equation:
$3\,\text{cosec}\,\theta = (\text{cosec}^2\theta - 1) - 17$
$18 + 3\,\text{cosec}\,\theta - \text{cosec}^2\theta = 0$ as required
[2 marks available — 1 mark for using correct identity, 1 mark for rearranging into required form]

(ii) To factorise the expression above, let $x = \text{cosec}\,\theta$.
Then $18 + 3x - x^2 = 0$, so $(6 - x)(3 + x) = 0$ ***[1 mark]***.
The roots of this quadratic occur at $x = 6$ and $x = -3$, so $\text{cosec}\,\theta = 6$ and $\text{cosec}\,\theta = -3$ ***[1 mark]***.
$\text{cosec}\,\theta = \frac{1}{\sin\theta}$, so $\sin\theta = \frac{1}{6}$ and $\sin\theta = -\frac{1}{3}$ ***[1 mark]***.
$\sin\theta = \frac{1}{6} \Rightarrow \theta = \sin^{-1}\frac{1}{6} = 0.16744...$ or
$\theta = \pi - 0.16744... = 2.97414...$
$\sin\theta = -\frac{1}{3} \Rightarrow \theta = \sin^{-1}\left(-\frac{1}{3}\right) = -0.33983...$
but this is outside the required range.
So $\theta = 2\pi + (-0.33983...) = 5.94334...$ or
$\theta = \pi - (-0.33983...) = 3.48142...$
So $\theta = 0.167, 2.97$ ***[1 mark]*** and
$\theta = 3.48, 5.94$ ***[1 mark]*** (all to 3 s.f.).

You don't have to use x = cosec θ — it's just a little easier to factorise without all those pesky cosecs flying around.

10 Using the small angle approximations, $\sin\theta \approx \theta$, $\cos\theta \approx 1 - \frac{1}{2}\theta^2$ and $\tan\theta \approx \theta$. Substituting these values into the expression gives:
$4\sin\theta\tan\theta + 2\cos\theta \approx 4(\theta \times \theta) + 2(1 - \frac{1}{2}\theta^2)$
$= 4\theta^2 + 2 - \theta^2 = 2 + 3\theta^2$ as required,
where $p = 2$ and $q = 3$.
[3 marks available — 1 mark for the correct approximations, 1 mark for substituting into the expression, 1 mark for rearranging to obtain the correct answer]

11 a) $\sec x = \frac{1}{\cos x}$, so as $\cos x = \frac{8}{9}$, $\sec x = \frac{9}{8}$ ***[1 mark]***.

b) The right-angled triangle with angle x, hypotenuse of length 9 and the adjacent side of length 8 (which gives the cos x value as stated) has the opposite side of length $\sqrt{9^2 - 8^2} = \sqrt{81 - 64} = \sqrt{17}$ ***[1 mark]***.
So the value of $\sin x = \frac{\sqrt{17}}{9}$ (opposite / hypotenuse).
$\text{cosec}\,x = \frac{1}{\sin x}$, so $\text{cosec}\,x = \frac{9}{\sqrt{17}} = \frac{9\sqrt{17}}{17}$ ***[1 mark]***.

c) For the triangle described in part b), the value of tan x is given by opposite / adjacent $= \frac{\sqrt{17}}{8}$ ***[1 mark]***.
So $\tan^2 x = \left(\frac{\sqrt{17}}{8}\right)^2 = \frac{17}{64}$ ***[1 mark]***.

You could have used $\sec^2 x \equiv 1 + \tan^2 x$ here instead.

d) $\cos 2x = 2\cos^2 x - 1$. Using the known value of $\cos x$,
$\cos 2x = 2\left(\frac{8}{9}\right)^2 - 1 = 2\left(\frac{64}{81}\right) - 1 = \frac{47}{81}$
[3 marks available — 1 mark for formula for cos 2x, 1 mark for working and 1 mark for correct answer]

You could have used the other versions of the cos 2x formula here ($\cos^2 x - \sin^2 x$ or $1 - 2\sin^2 x$) — just use the value you found for sin x in part b).

12 a) $\frac{1 + \cos x}{2} = \frac{1}{2}\left(1 + \cos 2\left(\frac{x}{2}\right)\right)$
$= \frac{1}{2}\left(1 + \left(2\cos^2\frac{x}{2} - 1\right)\right)$
$= \frac{1}{2}\left(2\cos^2\frac{x}{2}\right) = \cos^2\frac{x}{2}$
[2 marks available – 1 mark for using the correct identity, 1 mark for the correct rearrangement]

b) As $\cos^2\frac{x}{2} = 0.75$, $\frac{1 + \cos x}{2} = 0.75$.
So $1 + \cos x = 1.5$
$\cos x = 0.5$ ***[1 mark]*** $\Rightarrow x = \frac{\pi}{3}, \frac{5\pi}{3}$ ***[1 mark]***

You should know the solutions to cos x = 0.5 — it's one of the common angles.

13 $\sin 2\theta \equiv 2\sin\theta\cos\theta$, so $3\sin 2\theta\tan\theta \equiv 6\sin\theta\cos\theta\tan\theta$ ***[1 mark]***.
As $\tan\theta \equiv \frac{\sin\theta}{\cos\theta}$,
$6\sin\theta\cos\theta\tan\theta \equiv 6\sin\theta\cos\theta\frac{\sin\theta}{\cos\theta} \equiv 6\sin^2\theta$ ***[1 mark]***
so $3\sin 2\theta\tan\theta = 5 \Rightarrow 6\sin^2\theta = 5$ ***[1 mark]***
Then $\sin^2\theta = \frac{5}{6} \Rightarrow \sin\theta = \pm\sqrt{\frac{5}{6}} = \pm 0.9128...$ ***[1 mark]***
Solving this for θ gives $\theta = 1.15, 1.99, 4.29, 5.13$ ***[2 marks for all 4 correct answers, 1 mark for 2 correct answers]***

Don't forget the solutions for the negative square root as well — they're easy to miss. Drawing a sketch here is really useful — you can see that there are 4 solutions you need to find:

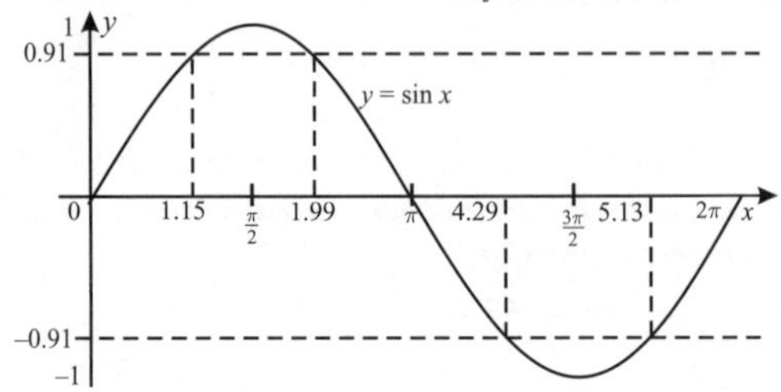

14 a) Use the double angle formula: $\cos 2\theta \equiv 1 - 2\sin^2\theta$ to replace $\cos 2\theta$:
$DE^2 = 4 - 4(1 - 2\sin^2\theta)$ ***[1 mark]***
$DE^2 = 4 - 4 + 8\sin^2\theta \Rightarrow DE^2 = 8\sin^2\theta$
$DE = \sqrt{8}\sin\theta = 2\sqrt{2}\sin\theta$ ***[1 mark for correct rearrangement]***

b) $P = 2DE + 2DG$
To find DG, use triangle BDG:

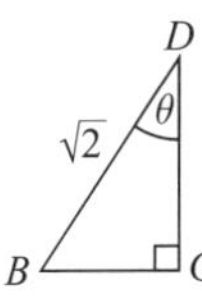

$\cos\theta = \frac{DG}{\sqrt{2}}$, so $DG = \sqrt{2}\cos\theta$ ***[1 mark]***
So $P = 2(2\sqrt{2}\sin\theta) + 2(\sqrt{2}\cos\theta)$
$= 4\sqrt{2}\sin\theta + 2\sqrt{2}\cos\theta$ ***[1 mark for correct substitution and rearrangement]***

c) $4\sqrt{2}\sin\theta + 2\sqrt{2}\cos\theta \equiv R\sin(\theta + \alpha)$
$\Rightarrow 4\sqrt{2}\sin\theta + 2\sqrt{2}\cos\theta \equiv R\sin\theta\cos\alpha + R\cos\theta\sin\alpha$
$\Rightarrow$ ① $R\cos\alpha = 4\sqrt{2}$ and ② $R\sin\alpha = 2\sqrt{2}$ ***[1 mark]***
② ÷ ① gives $\tan\alpha = \frac{1}{2} \Rightarrow \alpha = 0.464$ (3 s.f.) ***[1 mark]***
①² + ②² gives:
$R^2\cos^2\alpha + R^2\sin^2\alpha = (4\sqrt{2})^2 + (2\sqrt{2})^2 = 40$
$\Rightarrow R = \sqrt{40} = 2\sqrt{10}$ ***[1 mark]***
So $4\sqrt{2}\sin\theta + 2\sqrt{2}\cos\theta \equiv 2\sqrt{10}\sin(\theta + 0.464)$

15 a) $\sqrt{2}\cos\theta - 3\sin\theta \equiv R\cos(\theta + \alpha)$. Using the cos addition rule, $R\cos(\theta + \alpha) \equiv R\cos\theta\cos\alpha - R\sin\theta\sin\alpha$,
so ① $R\cos\alpha = \sqrt{2}$ and ② $R\sin\alpha = 3$ ***[1 mark]***.
② ÷ ① gives $\tan a = \frac{3}{\sqrt{2}} \Rightarrow \alpha = 1.13$ (3 s.f.) ***[1 mark]***
①² + ②² gives:
$R^2\cos^2\alpha + R^2\sin^2\alpha = (\sqrt{2})^2 + 3^2 = 11 \Rightarrow R = \sqrt{11}$ ***[1 mark]***,
so $\sqrt{2}\cos\theta - 3\sin\theta = \sqrt{11}\cos(\theta + 1.13)$.

b) The equation you're trying to solve is $\sqrt{2}\cos\theta - 3\sin\theta = 3$ in the range $0 \le \theta \le 6$, so, from part a), $\sqrt{11}\cos(\theta + 1.13) = 3$.
So $\cos(\theta + 1.13) = \frac{3}{\sqrt{11}}$.
Solving this gives $\theta + 1.13 = 0.4405...$ ***[1 mark]***.
The range of solutions becomes $1.13 \le \theta + 1.13 \le 7.13$.
To find the other values of θ within the new range, $2\pi - 0.4405...$
$= 5.842...$, $2\pi + 0.4405... = 6.723...$ ***[1 mark for both]***.
Subtracting 1.13 gives $\theta = 4.712...$, $5.593...$ mins ***[1 mark for both]***. So the water reaches 3 feet to the right of the sprinkler at 4 minutes
43 seconds and at 5 minutes 36 seconds ***[1 mark for both]***.
You can sketch the graph to help you find all the values of θ.

c) Using part a), $d = (\sqrt{2}\cos\theta - 3\sin\theta)^4 = (\sqrt{11}\cos(\theta + 1.13))^4$
The maximum distances left and right occur at the minimum and maximum value of d (the minimum value corresponds to the maximum distance left). The maximum and minimum points of $\cos(\theta + 1.13)$ are ± 1, so the maximum and minimum values of the function inside the brackets are $\pm\sqrt{11}$. This bracket is raised to the power 4, so the maximum distance right is: $(\pm\sqrt{11})^4 = 121$ feet ***[1 mark]***. Since $(\sqrt{11}\cos(\theta + 1.13))^4 \ge 0$, the maximum distance left is 0 feet ***[1 mark]***.
If you didn't realise that $(\sqrt{11}\cos(\theta + 1.13))^4$ is never negative, you'd have got the distance left wrong.

d) E.g. the sprinkler could be positioned against a wall, so it can never spray to the left ***[1 mark for a sensible comment]***.

16 $\text{cosec}\,2A \equiv \frac{1}{\sin 2A} \equiv \frac{1}{2\sin A\cos A}$

so $2\tan A\,\text{cosec}\,2A \equiv \frac{2\tan A}{2\sin A\cos A} \equiv \frac{2\frac{\sin A}{\cos A}}{2\sin A\cos A}$
$\equiv \frac{\sin A}{\sin A\cos^2 A} \equiv \frac{1}{\cos^2 A}$
$\equiv \sec^2 A \equiv 1 + \tan^2 A$

[3 marks available — 1 mark for using the double angle formula to expand sin 2A, 1 mark for rearranging and simplifying with the use of $\tan A \equiv \frac{\sin A}{\cos A}$, 1 mark for using $\sec^2 A \equiv 1 + \tan^2 A$ to get into the required form]

Pages 37-40: Exponentials and Logarithms

1 Rewrite all terms as powers of p and use the laws of logs to simplify:
$\log_p(p^4) + \log_p(p^{\frac{1}{2}}) - \log_p(p^{-\frac{1}{2}})$ ***[1 mark]***
$= 4\log_p p + \frac{1}{2}\log_p p - \left(-\frac{1}{2}\right)\log_p p$ ***[1 mark]***
$= 4 + \frac{1}{2} - \left(-\frac{1}{2}\right) = 4 + 1 = 5$ (as $\log_p p = 1$) ***[1 mark]***

2 $5^{(z^2-9)} = 2^{(z-3)}$, so taking logs of both sides gives:
$(z^2 - 9)\ln 5 = (z - 3)\ln 2$ ***[1 mark]***
$\Rightarrow (z + 3)(z - 3)\ln 5 - (z - 3)\ln 2 = 0$ ***[1 mark]***
$\Rightarrow (z - 3)[(z + 3)\ln 5 - \ln 2] = 0$ ***[1 mark]***
$\Rightarrow z - 3 = 0$ or $(z + 3)\ln 5 - \ln 2 = 0$
$\Rightarrow z = 3$ ***[1 mark]*** or $z = \frac{\ln 2}{\ln 5} - 3$
$\Rightarrow z = 3$ or $z = -2.57$ (3 s.f.) ***[1 mark]***

3 $3^{2x} = (3^x)^2$ (from the power laws), so let $y = 3^x$, then $y^2 = 3^{2x}$.
This gives a quadratic in y: $y^2 - 9y + 14 = 0$
$(y - 2)(y - 7) = 0$ ***[1 mark]***, so $y = 2$ or $y = 7$
$\Rightarrow 3^x = 2$ or $3^x = 7$ ***[1 mark for both]***
To solve these equations, take logs of both sides ***[1 mark]***.
$3^x = 2 \Rightarrow \log 3^x = \log 2 \Rightarrow x\log 3 = \log 2$
$\Rightarrow x = \frac{\log 2}{\log 3} = 0.631$ (3 s.f.) ***[1 mark]***
$3^x = 7 \Rightarrow \log 3^x = \log 7 \Rightarrow x\log 3 = \log 7$
$\Rightarrow x = \frac{\log 7}{\log 3} = 1.77$ (3 s.f.) ***[1 mark]***

4 a) $y = \ln(4x - 3)$, and $x = a$ when $y = 1$.
$1 = \ln(4a - 3) \Rightarrow e^1 = 4a - 3$ ***[1 mark]***
$\Rightarrow a = \frac{e^1 + 3}{4} = 1.43$ to 2 d.p. ***[1 mark]***

b) The curve can only exist when $4x - 3 > 0$ ***[1 mark]***
$\Rightarrow x > \frac{3}{4} \Rightarrow x > 0.75$, so $b = 0.75$ ***[1 mark]***.

5 a) A is the value of y when $x = 0$, so $A = 4$ ***[1 mark]***.
Now use this value to find b:
$\frac{4}{e} = 4e^{10b}$ ***[1 mark]*** $\Rightarrow 4e^{-1} = 4e^{10b}$
$\Rightarrow -1 - 10b \rightarrow b = -0.1$ ***[1 mark]***

b) The gradient of $y = Ae^{bx}$ is bAe^{bx}. Here, $A = 4$ and $b = -0.1$, so the gradient is $-0.1 \times 4 \times e^{-0.1x} = -0.4e^{-0.1x}$ ***[1 mark for a correct gradient expression]***.
Set this equal to the value given and solve:
$-0.4e^{-0.1x} = -1$ ***[1 mark]*** $\Rightarrow e^{-0.1x} = 2.5$
$\Rightarrow -0.1x = \ln 2.5$ ***[1 mark]*** $\Rightarrow x = -10\ln 2.5$ ***[1 mark]***
When $x = -10\ln 2.5$, $y = 4e^{-0.1(-10\ln 2.5)} = 4e^{\ln 2.5} = 4 \times 2.5 = 10$
So the exact coordinates are $(-10\ln 2.5, 10)$ ***[1 mark]***.

6 $y = e^{ax} + b$
The sketch shows that when $x = 0$, $y = -6$, so:
$-6 = e^0 + b \Rightarrow -6 = 1 + b \Rightarrow b = -7$ ***[1 mark]***.
The sketch also shows that when $y = 0$, $x = \frac{1}{4}\ln 7$, so:
$0 = e^{(\frac{a}{4}\ln 7)} - 7$ ***[1 mark]*** $\Rightarrow e^{(\frac{a}{4}\ln 7)} = 7$
$\Rightarrow \frac{a}{4}\ln 7 = \ln 7 \Rightarrow \frac{a}{4} = 1 \Rightarrow a = 4$ ***[1 mark]***.
The asymptote occurs as $x \to -\infty$, so $e^{4x} \to 0$,
and since $y = e^{4x} - 7$, $y \to -7$.
So the equation of the asymptote is $y = -7$ ***[1 mark]***.
You could also have solved this question by thinking about the series of transformations that would take you from the graph of $y = e^x$ to this one.

7 The value after the first year is $0.92 \times 8000 = £7360$ ***[1 mark]***.
You need to find n, the number of months after the first year when the value falls below £4000. So solve the equation:
$7360 \times e^{\frac{n}{12}\ln\left(1-\frac{4}{100}\right)} = 4000$ ***[1 mark]*** $\Rightarrow e^{\frac{n}{12}\ln 0.96} = \frac{4000}{7360}$
$\Rightarrow \frac{n}{12}\ln 0.96 = \ln\left(\frac{4000}{7360}\right)$ ***[1 mark]***
$\Rightarrow n = 12\ln\frac{4000}{7360} \div \ln 0.96 = 179.246$ ***[1 mark]***
179 months after the first year the value is £4003.35 to the nearest penny (i.e. > 4000), so you need to round up to 180 months.
The total number of months is: $12 + 180 = 192$ months ***[1 mark]***.
Don't forget to add the 12 at the end — that's the months from the first year (which had a different rate of depreciation).

8 a) You need to find t such that:
$2100 - 1500e^{-0.15t} > 5700e^{-0.15t}$ ***[1 mark]***
$2100 > 7200e^{-0.15t}$
$\frac{7}{24} > e^{-0.15t}$
$\ln\frac{7}{24} > -0.15t$ ***[1 mark]***
$\ln\frac{7}{24} \div -0.15 < t$
$t > 8.21429...$ ***[1 mark]***
So the population of Q first exceeds the population of P when $t > 8.21$ (3 s.f.), i.e. in the year 2018 ***[1 mark]***.
Don't forget to flip the inequality sign when you divide by −0.15.

b)
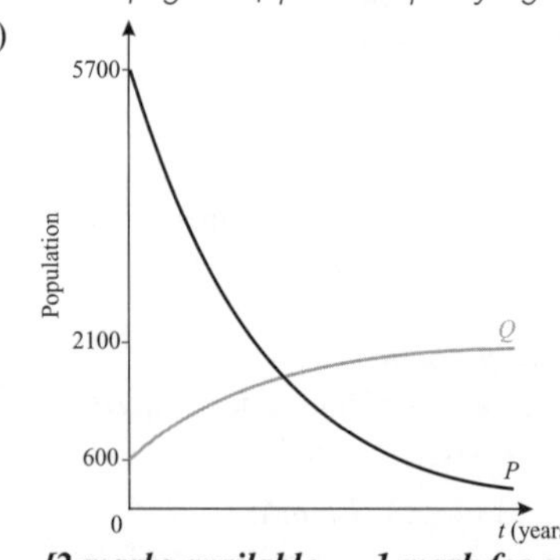

[2 marks available — 1 mark for correct shape of graph, 1 mark for (0, 5700) labelled]

c) Bird of prey — e.g. any one of:
- The model predicts the population of the birds of prey will increase, but will tend to a limit. This seems realistic, as the bird of prey will have to compete for the available sources of food as one source decreases.
- The population grows quite slowly (especially compared to the rate of decrease of the other species) — this seems more realistic than a rapid population growth.
- The rate of growth slows over time, which would be expected as food supplies dwindle.

[1 mark for a sensible comment about the birds of prey]
Endangered species — e.g. any one of:
- The model predicts the population will decrease, which seems realistic as the birds of prey will hunt them.
- The model predicts a very rapid decline at first, which does not seem realistic — you'd expect the rate of decrease to be slower at first.

[1 mark for a sensible comment about the endangered species]

d) You need to find t such that $5700e^{-0.15t} = 1000$ ***[1 mark]***
$\Rightarrow 1000 = \frac{5700}{e^{0.15t}} \Rightarrow e^{0.15t} = \frac{5700}{1000} = 5.7.$
$\Rightarrow 0.15t = \ln 5.7 \Rightarrow t = \frac{\ln 5.7}{0.15} = 11.6031...$ years ***[1 mark]***
So the population is predicted to drop below 1000 in the year 2021 ***[1 mark]***.
If you set up and solved an inequality that's fine — you'd still get the marks.

e) E.g. The function could be refined so that from 2021, the population is predicted to stop decreasing — it could either level out or start increasing ***[1 mark for a sensible comment]***.

9 a) $y = ab^t$, so take logs of both sides: $\log y = \log ab^t$
Then use the laws of logs: $\log y = \log a + \log b^t$ ***[1 mark]***
$\log y = \log a + t\log b$ ***[1 mark]***
$\log y = t\log b + \log a$, as required.

b) First find the values of a and b:
Comparing $\log y = t\log b + \log a$ to $y = mx + c$ gives $\log b = m$, the gradient of the graph, and $\log a = c$, the vertical-axis intercept of the graph.
Use points from the graph to calculate the gradient, m:
For example, using the points (2, 0.3) and (1, 0):
$m = \frac{y_2 - y_1}{x_2 - x_1} = \frac{0.3 - 0}{2 - 1} = 0.3$
So $\log b = 0.3 \Rightarrow b = 10^{0.3}$ ***[1 mark]***
Now estimate the vertical-axis intercept to find $\log_{10} a$:
$\log a = -0.3 \Rightarrow a = 10^{-0.3}$ ***[1 mark]***.
The equation is $y = 10^{-0.3} \times (10^{0.3})^t = 10^{0.3t - 0.3} = 10^{0.3(t-1)}$
y is the average attendance in hundreds, so the attendance exceeds 5000 when $y > 50$, i.e. $10^{0.3(t-1)} > 50$ ***[1 mark]***
$0.3(t - 1) > \log 50 \Rightarrow t > \frac{\log 50}{0.3} + 1$
$\Rightarrow t > 6.663...$ years after the 2010/2011 season ***[1 mark]***.
This is during the 2016/2017 season ***[1 mark]***.
Don't worry if your values of a and b are slightly different — you should end up with the same answer though.

c) a is the average attendance in hundreds in the season where $t = 0$, i.e in the 2010/11 season ***[1 mark]***.
The attendance was around 50 supporters.

d) For the 2024/25 season, $t = 14$, which is beyond the values of t given on the graph ***[1 mark]***. This is extrapolation, so may not be accurate as the model might not hold that far in the future ***[1 mark]***.

Pages 41-44: Differentiation 1

1 The gradient of the tangent is the same as the gradient of the curve, so differentiate:
$\frac{dy}{dx} = 6x^2 - 20x - 2x^{-\frac{1}{2}} = 6x^2 - 20x - \frac{2}{\sqrt{x}}$
Now put $x = 4$ into your derivative:
$6(4^2) - 20(4) - \frac{2}{\sqrt{4}} = 96 - 80 - 1 = 15$
[4 marks available — 1 mark for differentiating, 1 mark for the correct derivative, 1 mark for substituting in x = 4, 1 mark for the correct answer]

2 Differentiate f(x) to find f'(x):
$f'(x) = 3x^2 - 14x + 8$
So the graph of f'(x) is a positive quadratic (i.e. u-shaped).
It crosses the y-axis when $x = 0$, which gives a y-value of 8.
It crosses the x-axis when $3x^2 - 14x + 8 = 0$
$\Rightarrow (3x - 2)(x - 4) = 0$, so $x = \frac{2}{3}$ and $x = 4$.
Now sketch the graph:

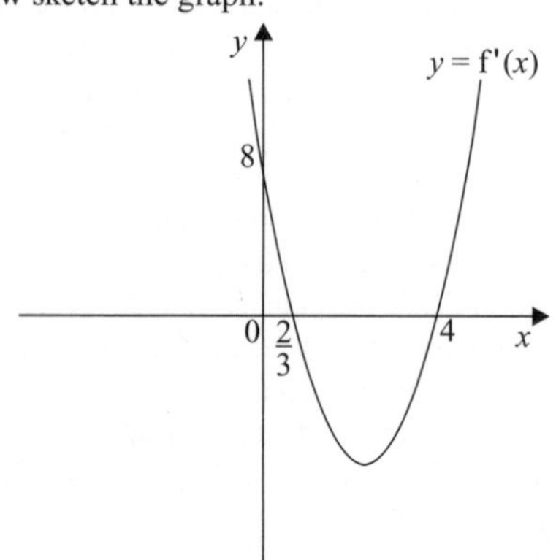

[4 marks available — 1 mark for attempting to differentiate, 1 mark for the correct function for y = f'(x), 1 mark for the correct shape of the graph, 1 mark for the correct x- and y-intercepts]

3 a) The gradient of the normal at R is the same as the gradient of the line $4y + x = 24$. Rearrange this equation to find the gradient:

$4y + x = 24 \Rightarrow 4y = 24 - x \Rightarrow y = 6 - \frac{1}{4}x$,

so gradient of normal at R $= -\frac{1}{4}$

Gradient of curve at R $= -1 \div -\frac{1}{4} = 4$

Find an expression for the gradient of the curve by differentiating $y = kx^2 - 8x - 5$:

$\frac{dy}{dx} = 2kx - 8$

At R, gradient $= 2k(2) - 8 = 4k - 8$

Put this expression equal to the value of the gradient at R to find k:

$4k - 8 = 4 \Rightarrow 4k = 12 \Rightarrow k = 3$

[5 marks available — 1 mark for finding the gradient of the normal at R, 1 mark for finding the gradient of the curve at R, 1 mark for attempting to differentiate y, 1 mark for forming an equation for k, 1 mark for the correct value of k]

b) Gradient of tangent at R = gradient of curve = 4.

At R, $x = 2$, so $y = 3(2^2) - 8(2) - 5 = -9$ ***[1 mark]***

Use these values in $y - y_1 = m(x - x_1)$ to find the equation of the tangent:

$y + 9 = 4(x - 2) \Rightarrow y + 9 = 4x - 8 \Rightarrow y = 4x - 17$ ***[1 mark]***

Equate $y = 4x - 17$ and $y = 4x - \frac{1}{x^3} - 9$ to find S:

$4x - 17 = 4x - \frac{1}{x^3} - 9$ ***[1 mark]***

$-8 = -\frac{1}{x^3} \Rightarrow x = \frac{1}{2}$ and $y = 4(\frac{1}{2}) - 17 = -15$

So at S, $x = \frac{1}{2}$ ***[1 mark]*** and $y = -15$ ***[1 mark]***

4 $f'(x) = \lim_{h \to 0}\left(\frac{(8(x+h)^2 - 1) - (8x^2 - 1)}{h}\right)$ ***[1 mark]***

$= \lim_{h \to 0}\left(\frac{(8(x^2 + 2xh + h^2) - 1) - (8x^2 - 1)}{h}\right)$

$= \lim_{h \to 0}\left(\frac{8x^2 + 16xh + 8h^2 - 1 - 8x^2 + 1}{h}\right)$ ***[1 mark]***

$= \lim_{h \to 0}\left(\frac{16xh + 8h^2}{h}\right)$

$= \lim_{h \to 0}(16x + 8h)$ ***[1 mark]***

As $h \to 0$, $16x + 8h \to 16x$, so $f'(x) = 16x$

[1 mark for letting h → 0 and obtaining the correct limit]

5 a) Differentiate f(x) and set the derivative equal to zero:

$f'(x) = 8x^3 + 27$ ***[1 mark]***

$8x^3 + 27 = 0$ ***[1 mark]*** $\Rightarrow x^3 = -\frac{27}{8}$

$\Rightarrow x = \sqrt[3]{-\frac{27}{8}} = -\frac{3}{2} = -1.5$ ***[1 mark]***

When $x = -1.5$, $f(x) = 2(-1.5)^4 + 27(-1.5) = -30.375$ ***[1 mark]***

So the stationary point is at (–1.5, –30.375)

b) The function is increasing if the gradient is positive.

$f'(x) > 0$ if $8x^3 + 27 > 0 \Rightarrow x^3 > -\frac{27}{8} \Rightarrow x > \sqrt[3]{-\frac{27}{8}}$

$\Rightarrow x > -1.5$

The function is decreasing if the gradient is negative.

$f'(x) < 0$ if $8x^3 + 27 < 0 \Rightarrow x^3 < -\frac{27}{8} \Rightarrow x < \sqrt[3]{-\frac{27}{8}}$

$\Rightarrow x < -1.5$

[2 marks available — 1 mark for forming at least one correct inequality, 1 mark for both ranges of values correct]

c) You know from parts a) and b) that the function has a stationary point at (–1.5, –30.375) and that this is a minimum point because the function is decreasing to the left of this point and increasing to the right of it.

Find where the curve crosses the y-axis:

When $x = 0$, $f(x) = 0$, so the curve goes through the origin.

Find where the curve crosses the x-axis:

When $f(x) = 0$, $2x^4 + 27x = 0 \Rightarrow x(2x^3 + 27) = 0$

$\Rightarrow x = 0$ or $2x^3 + 27 = 0$

$\Rightarrow x^3 = -\frac{27}{2} \Rightarrow x = \sqrt[3]{-\frac{27}{2}} = -2.381$ (3 d.p.),

so the curve crosses the x-axis at $x = 0$ and $x = -2.381$.

Now use the information you've found to sketch the curve:

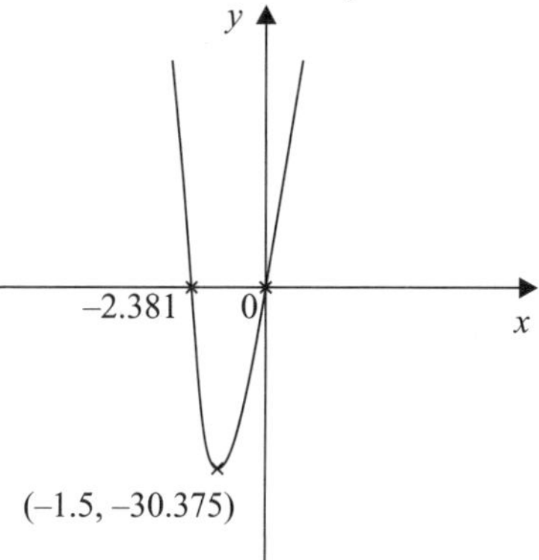

[3 marks available — 1 mark for a curve with the correct shape, 1 mark for the correct minimum point, 1 mark for the correct intercepts]

6 The graph is concave downwards when $\frac{d^2y}{dx^2} < 0$.

$y = x^4 + 3x^3 - 6x^2 \Rightarrow \frac{dy}{dx} = 4x^3 + 9x^2 - 12x$ ***[1 mark]***

$\Rightarrow \frac{d^2y}{dx^2} = 12x^2 + 18x - 12$ ***[1 mark]***

$12x^2 + 18x - 12 < 0$ ***[1 mark]*** $\Rightarrow 2x^2 + 3x - 2 < 0$

$\Rightarrow (2x - 1)(x + 2) < 0$

$(2x - 1)(x + 2) < 0$ for $x < \frac{1}{2}$ and $x > -2$,

so the graph is concave downwards for $-2 < x < \frac{1}{2}$ ***[1 mark]***.

To check your inequality think about the shape of the graph of $\frac{d^2y}{dx^2}$ — it's a positive quadratic, so is less than 0 between –2 and $\frac{1}{2}$.

7 a) At a point of inflection, $f''(x) = 0$, so find $f''(x)$.

$f(x) = 3x^3 + 9x^2 + 25x \Rightarrow f'(x) = 9x^2 + 18x + 25$ ***[1 mark]***

$\Rightarrow f''(x) = 18x + 18$ ***[1 mark]***

$f''(-1) = 18(-1) + 18 = 0$, so $f''(x) = 0$ at $x = -1$ ***[1 mark]***

To confirm that this is a point of inflection, you need to check what's happening either side of $x = -1$:

For $x < -1$, $f''(x) < 0$ and for $x > -1$, $f''(x) > 0$ ***[1 mark]***.

The curve changes from concave downwards to concave upwards, so $x = -1$ is a point of inflection ***[1 mark]***.

b) At a stationary point, $f'(x) = 0$.

$f'(x) = 9x^2 + 18x + 25$

$f'(-1) = 9(-1)^2 + 18(-1) + 25 = 16$.

Since $16 \neq 0$, this is a non-stationary point of inflection.

[2 marks available — 1 mark for finding f'(–1), 1 mark for a correct explanation of why this is a non-stationary point of inflection]

c) f(x) is an increasing function for all values of x if $f'(x) > 0$ for all x.

From a), $f'(x) = 9x^2 + 18x + 25$

Complete the square to show that $f'(x) > 0$:

$f'(x) = 9x^2 + 18x + 25$

$\Rightarrow f'(x) = 9(x^2 + 2x) + 25$

$\Rightarrow f'(x) = 9(x + 1)^2 - 9 + 25$

$\Rightarrow f'(x) = 9(x + 1)^2 + 16$ ***[1 mark]***

$(x + 1)^2 \geq 0$, so $f'(x)$ has a minimum value of 16

So $f'(x) > 0$ for all x, which means that f(x) is an increasing function for all values of x ***[1 mark]***.

8 a) Surface area of the container = sum of the areas of all 5 faces $= x^2 + x^2 + xy + xy + xy = 2x^2 + 3xy$ ***[1 mark]***

40 litres = 40 000 cm^3

Volume of the container = length × width × height

$= x^2y = 40\,000$ cm^3 ***[1 mark]*** $\Rightarrow y = \frac{40\,000}{x^2}$

Put this into the formula for the area:

$A = 2x^2 + 3xy = 2x^2 + 3x\left(\frac{40\,000}{x^2}\right)$ ***[1 mark]***

$= 2x^2 + \frac{120\,000}{x}$ ***[1 mark]***

b) To find stationary points, first find $\frac{dA}{dx}$:

$\frac{dA}{dx} = 4x - \frac{120\,000}{x^2}$ ***[1 mark for attempting to differentiate, 1 mark for the correct function]***

Then find the value of x where $\frac{dA}{dx} = 0$:

$4x - \frac{120\,000}{x^2} = 0$ ***[1 mark]*** $\Rightarrow x^3 = 30\,000$

$\Rightarrow x = 31.07... = 31.1$ cm (3 s.f.) ***[1 mark]***

To check if it's a minimum, find $\frac{d^2A}{dx^2}$:

$\frac{d^2A}{dx^2} = 4 + \frac{240\,000}{x^3} = 12$ at $x = 31.07...$ ***[1 mark]***.

The second derivative is positive, so it's a minimum ***[1 mark]***

c) Put the value of x found in part b) into the formula for the area given in part a):

$A = 2(31.07...)^2 + \frac{120\,000}{31.07...}$ ***[1 mark]*** $= 5792.936...$

$= 5790$ cm^2 (3 s.f.) ***[1 mark]***

d) E.g. the model does not take into account the thickness of the steel, so the minimum area needed is likely to be slightly greater than this to create the required capacity ***[1 mark for a sensible comment]***.

Pages 45-50: Differentiation 2

1 a) $y = \frac{1}{\sqrt{2x - x^2}} = (2x - x^2)^{-\frac{1}{2}}$

Let $u = 2x - x^2$, so $y = u^{-\frac{1}{2}}$ ***[1 mark]***,

then $\frac{du}{dx} = 2 - 2x$ and $\frac{dy}{du} = -\frac{1}{2}u^{-\frac{3}{2}}$

Using the chain rule: $\frac{dy}{dx} = \frac{dy}{du} \times \frac{du}{dx} = -\frac{1}{2}u^{-\frac{3}{2}} \times (2 - 2x)$

$= -\frac{1}{2}(2x - x^2)^{-\frac{3}{2}} \times (2 - 2x)$ ***[1 mark]***

$= -\frac{2 - 2x}{2(\sqrt{2x - x^2})^3} = \frac{x - 1}{(\sqrt{2x - x^2})^3}$ ***[1 mark]***

So at (1, 1), $\frac{dy}{dx} = 0$ ***[1 mark]***.

b) $x = (4y + 10)^3$

Let $u = 4y + 10$, so $x = u^3$ ***[1 mark]***,

then $\frac{du}{dy} = 4$ and $\frac{dx}{du} = 3u^2$

Using the chain rule: $\frac{dx}{dy} = \frac{dx}{du} \times \frac{du}{dy} = 3u^2 \times 4$

$= 12(4y + 10)^2$ ***[1 mark]***

So $\frac{dy}{dx} = \frac{1}{\left(\frac{dx}{dy}\right)} = \frac{1}{12(4y + 10)^2}$ ***[1 mark]***

So at (8, –2), $\frac{dy}{dx} = \frac{1}{48}$ ***[1 mark]***

You could have found the answer by rearranging the equation to get y on its own and then differentiating normally.

2 a) Replace h with x in the height formula:

$x = \sqrt{\frac{2}{3}}a \Rightarrow a = \sqrt{\frac{3}{2}}x$ ***[1 mark]***

Now substitute $a = \sqrt{\frac{3}{2}}x$ into the expression for volume:

$V = \frac{\sqrt{2}}{12}\left(\sqrt{\frac{3}{2}}x\right)^3 = \frac{\sqrt{2}}{12} \times \frac{3\sqrt{3}}{2\sqrt{2}}x^3 = \frac{\sqrt{3}}{8}x^3$

[1 mark for substitution and correct simplification]

b) From the question, you know that the rate of change of volume with respect to time, $\frac{dV}{dt}$, is 240. And differentiating the expression for volume from part a) with respect to x gives

$\frac{dV}{dx} = \frac{3\sqrt{3}x^2}{8}$ ***[1 mark]***

Using the chain rule: $\frac{dx}{dt} = \frac{dx}{dV} \times \frac{dV}{dt}$ ***[1 mark]***

$= \frac{1}{\left(\frac{dV}{dx}\right)} \times \frac{dV}{dt} = \frac{8}{3\sqrt{3}x^2} \times 240 = \frac{640}{\sqrt{3}x^2}$ ***[1 mark]***

So when $x = 8$, $\frac{dx}{dt} = \frac{640}{64\sqrt{3}} = \frac{10}{\sqrt{3}} = \frac{10\sqrt{3}}{3}$ cm min^{-1}

[1 mark for substitution of x = 8, 1 mark for correct answer in surd form with correct units]

c) $\frac{dV}{dt} = \frac{dV}{dx} \times \frac{dx}{dt}$ ***[1 mark]***

$= \frac{3\sqrt{3}x^2}{8} \times \frac{32}{9\sqrt{3}} = \frac{4x^2}{3}$ ***[1 mark]***

So when $x = 12$, $\frac{dV}{dt} = \frac{4 \times 144}{3} = 192$ cm^3 min^{-1} ***[1 mark]***

3 a) $y = e^{2x} - 5e^x + 3x$, so using chain rule:

$\frac{dy}{dx} = 2e^{2x} - 5e^x + 3$

[1 mark for $2e^{2x}$, 1 mark for the other two terms correct]

b) Stationary points occur when $\frac{dy}{dx} = 0$, so:

$2e^{2x} - 5e^x + 3 = 0$ ***[1 mark]***.

This looks like a quadratic, so substitute $z = e^x$ and factorise:

$2z^2 - 5z + 3 = 0 \Rightarrow (2z - 3)(z - 1) = 0$ ***[1 mark]***

So the solutions are:

$2z - 3 = 0 \Rightarrow z = \frac{3}{2} \Rightarrow e^x = \frac{3}{2} \Rightarrow x = \ln\frac{3}{2}$ ***[1 mark]***, and

$z - 1 = 0 \Rightarrow z = 1 \Rightarrow e^x = 1 \Rightarrow x = \ln 1 = 0$ ***[1 mark]***.

c) To determine the nature of the stationary points, find $\frac{d^2y}{dx^2}$ at $x = 0$ and $x = \ln\frac{3}{2}$:

$\frac{d^2y}{dx^2} = 4e^{2x} - 5e^x$ ***[1 mark]***

When $x = 0$, $\frac{d^2y}{dx^2} = 4e^0 - 5e^0 = 4 - 5 = -1$,

so $\frac{d^2y}{dx^2} < 0$, which means the point is a maximum ***[1 mark]***.

When $x = \ln\frac{3}{2}$, $\frac{d^2y}{dx^2} = 4e^{2\ln\frac{3}{2}} - 5e^{\ln\frac{3}{2}} = 4\left(\frac{3}{2}\right)^2 - 5\left(\frac{3}{2}\right) = \frac{3}{2}$,

so $\frac{d^2y}{dx^2} > 0$, which means the point is a minimum ***[1 mark]***.

4 a) Use the chain rule to find $\frac{dy}{dx}$:

$y = 2^{3x}$, so $u = 3x$ and $y = 2^u$, ***[1 mark]***

so $\frac{du}{dx} = 3$ and $\frac{dy}{du} = 2^u \ln 2$

$\Rightarrow \frac{dy}{dx} = 2^u \ln 2 \times 3 = 3(2^{3x} \ln 2)$ ***[1 mark]***

At $x = 1$, $\frac{dy}{dx} = 3(2^3 \ln 2) = 24 \ln 2$ ***[1 mark]***

b) y is an increasing function for all values of x if $\frac{dy}{dx} > 0$ for all x ***[1 mark]***. From a), $\frac{dy}{dx} = 3(2^{3x} \ln 2)$.

Since $2^{3x} > 0$ for all x and 3 and ln 2 are both positive, $3(2^{3x} \ln 2) > 0$ for all x ***[1 mark]***.

5 $f(x) = \cos x$, so:

$f'\left(\frac{\pi}{3}\right) = \lim_{h\to 0}\left(\frac{\cos\left(\frac{\pi}{3} + h\right) - \cos\left(\frac{\pi}{3}\right)}{h}\right)$ ***[1 mark]***

$= \lim_{h\to 0}\left(\frac{\cos\frac{\pi}{3}\cos h - \sin\frac{\pi}{3}\sin h - \cos\left(\frac{\pi}{3}\right)}{h}\right)$ ***[1 mark]***

$= \lim_{h\to 0}\left(\frac{\cos\frac{\pi}{3}(\cos h - 1) - \sin\frac{\pi}{3}\sin h}{h}\right)$

$= \lim_{h\to 0}\left(\frac{\frac{1}{2}\left(-\frac{1}{2}h^2\right) - \frac{\sqrt{3}}{2}h}{h}\right)$

[1 mark for using small angle approximations for each of cos h and sin h]

$= \lim_{h\to 0}\left(-\frac{h}{4} - \frac{\sqrt{3}}{2}\right)$ ***[1 mark]***

As $h \to 0$, $-\frac{h}{4} \to 0$, so $f'\left(\frac{\pi}{3}\right) \to -\frac{\sqrt{3}}{2}$ ***[1 mark]***

6 $y = \ln x\,(5x - 2)^3$, so use the product rule with $u = \ln x$ and $v = (5x - 2)^3$.

First use the chain rule to find $\frac{dv}{dx} = 15(5x - 2)^2$.

So $u = \ln x$, $\frac{du}{dx} = \frac{1}{x}$, $v = (5x - 2)^3$, $\frac{dv}{dx} = 15(5x - 2)^2$

Then $\frac{dy}{dx} = u\frac{dv}{dx} + v\frac{du}{dx} = \ln x \times 15(5x - 2)^2 + (5x - 2)^3 \times \frac{1}{x}$

$= 15 \ln x\,(5x - 2)^2 + \frac{(5x - 2)^3}{x}$

$= (5x - 2)^2\left[15\ln x + \frac{(5x - 2)}{x}\right]$

$= (5x - 2)^2\left[15\ln x + 5 - \frac{2}{x}\right]$

So $a = 15$, $b = 5$ and $c = -2$.

[4 marks available — 1 mark each for finding expressions for du/dx and dv/dx, 1 mark for putting these expressions into the product rule, 1 mark for the answer in the correct form with correct a, b and c]

7 Use the quotient rule to find $\frac{dy}{dx}$:

$u = 5 - 2x \Rightarrow \frac{du}{dx} = -2$

$v = 3x^2 + 3x \Rightarrow \frac{dv}{dx} = 6x + 3$

$$\frac{dy}{dx} = \frac{v\frac{du}{dx} - u\frac{dv}{dx}}{v^2}$$

$$= \frac{(3x^2 + 3x)(-2) - (5 - 2x)(6x + 3)}{(3x^2 + 3x)^2}$$

$$= \frac{-6x^2 - 6x - (30x + 15 - 12x^2 - 6x)}{(3x^2 + 3x)^2} = \frac{6x^2 - 30x - 15}{(3x^2 + 3x)^2}$$

At a stationary point, $\frac{dy}{dx} = 0$.

$\frac{6x^2 - 30x - 15}{(3x^2 + 3x)^2} = 0 \Rightarrow 6x^2 - 30x - 15 = 0 \Rightarrow 2x^2 - 10x - 5 = 0$

[6 marks available — 1 mark each for du/dx and dv/dx, 1 mark for putting these expressions into the quotient rule to find dy/dx, 1 mark for correct expression for dy/dx, 1 mark for setting dy/dx equal to zero, 1 mark for simplifying to show that $2x^2 - 10x - 5 = 0$]

8 $y = 4x^2 \ln x$, so use the product rule to find $\frac{dy}{dx}$:

$u = 4x^2$ and $v = \ln x$, so $\frac{du}{dx} = 8x$ and $\frac{dv}{dx} = \frac{1}{x}$ ***[1 mark]***

So $\frac{dy}{dx} = 4x^2 \times \frac{1}{x} + \ln x \times 8x$

$= 4x + 8x \ln x$ ***[1 mark]***

Now use the product rule with $u = 8x$ and $v = \ln x$ to find $\frac{d^2y}{dx^2}$:

$u = 8x$ and $v = \ln x$, so $\frac{du}{dx} = 8$ and $\frac{dv}{dx} = \frac{1}{x}$ ***[1 mark]***

So $\frac{d^2y}{dx^2} = 4 + \left(8x \times \frac{1}{x} + \ln x \times 8\right)$

$= 4 + 8 + 8 \ln x$

$= 12 + 8 \ln x$ ***[1 mark]***

The curve is concave downwards for $\frac{d^2y}{dx^2} < 0$.

$12 + 8 \ln x < 0$ ***[1 mark]*** $\Rightarrow \ln x < -1.5 \Rightarrow x < e^{-1.5}$

You know that $x > 0$, so the curve is concave downwards for $0 < x < e^{-1.5}$ ***[1 mark]***.

The curve is concave upwards for $\frac{d^2y}{dx^2} > 0$.

$12 + 8 \ln x > 0$ ***[1 mark]*** $\Rightarrow \ln x > -1.5 \Rightarrow x > e^{-1.5}$, so the curve is concave upwards for $x > e^{-1.5}$ ***[1 mark]***.

You could have factorised $\frac{dy}{dx}$ to get $4x(1 + 2 \ln x)$. This would make the following few steps a bit different, but the answer will be the same.

9 a) $y = \frac{4x - 1}{\tan x}$, so use the quotient rule:

$u = 4x - 1 \Rightarrow \frac{du}{dx} = 4$

$v = \tan x \Rightarrow \frac{dv}{dx} = \sec^2 x$

$$\frac{dy}{dx} = \frac{v\frac{du}{dx} - u\frac{dv}{dx}}{v^2} = \frac{4\tan x - (4x - 1)\sec^2 x}{\tan^2 x}$$

$$= \frac{4}{\tan x} - \frac{(4x - 1)\sec^2 x}{\tan^2 x}$$

Since $\frac{1}{\tan x} = \cot x$, $\sec^2 x = \frac{1}{\cos^2 x}$ and $\tan^2 x = \frac{\sin^2 x}{\cos^2 x}$:

$$\frac{dy}{dx} = 4 \cot x - \frac{(4x - 1)}{\cos^2 x\left(\frac{\sin^2 x}{\cos^2 x}\right)} = 4 \cot x - \frac{(4x - 1)}{\sin^2 x}$$

Since $\frac{1}{\sin^2 x} = \text{cosec}^2 x$:

$\frac{dy}{dx} = 4 \cot x - (4x - 1) \text{cosec}^2 x$

[3 marks available — 1 mark for correct expressions for du/dx and dv/dx, 1 mark for correct use of the quotient rule, and 1 mark for reaching the correct expression for dy/dx]

Alternatively, you could have used the product rule with $y = (4x - 1)\cot x$.

b) Maximum point is when $\frac{dy}{dx} = 0$:

$4 \cot x - (4x - 1) \text{cosec}^2 x = 0$ ***[1 mark]***

Dividing through by $\text{cosec}^2 x$ gives

$\frac{4\cot x}{\text{cosec}^2 x} - (4x - 1) = 0$ ***[1 mark]***

$\frac{4\cot x}{\text{cosec}^2 x} = \frac{4\cos x \sin^2 x}{\sin x} = 4\cos x \sin x$ ***[1 mark]***,

and using the double angle formula, $\sin 2x = 2 \sin x \cos x$, so $4 \cos x \sin x = 2 \sin 2x$.

So $2 \sin 2x - 4x + 1 = 0$ ***[1 mark]***.

You're told that the point is a maximum, so you don't need to differentiate again to check.

10 $\frac{dx}{dt} = 2t$, $\frac{dy}{dt} = 3t^2 + 2$

So $\frac{dy}{dx} = \frac{dy}{dt} \div \frac{dx}{dt} = \frac{3t^2 + 2}{2t}$

Substitute $t = 1$ into this expression to find the gradient of the curve at $t = 1$:

$\frac{dy}{dx} = \frac{3(1)^2 + 2}{2(1)} = \frac{5}{2}$

Now find the coordinates of the point where $t = 1$:

$x = 1^2 + 1 = 2$ and $y = 1^3 + 2(1) = 3$

Now use the gradient at $t = 1$ and the point (2, 3) to find the equation of the tangent:

$y - 3 = \frac{5}{2}(x - 2) \Rightarrow y - 3 = \frac{5}{2}x - 5 \Rightarrow y = \frac{5}{2}x - 2$

[5 marks available — 1 mark for a correct method to find dy/dx, 1 mark for a correct expression for dy/dx, 1 mark for substituting t = 1 to find the gradient, 1 mark for finding the coordinates when t = 1, 1 mark for the correct equation]

11 $y = \cos^{-1} x \Rightarrow \cos y = x$ ***[1 mark]***

Differentiate with respect to x:

$-\sin y \times \frac{dy}{dx} = 1 \Rightarrow \frac{dy}{dx} = -\frac{1}{\sin y}$ ***[1 mark]***

Using the identity $\cos^2 y + \sin^2 y = 1$,

$\sin^2 y = 1 - \cos^2 y \Rightarrow \sin y = \sqrt{1 - \cos^2 y}$

So $\frac{dy}{dx} = -\frac{1}{\sqrt{1 - \cos^2 y}}$ ***[1 mark]***

As $\cos y = x$, this expression becomes $\frac{dy}{dx} = -\frac{1}{\sqrt{1 - x^2}}$ ***[1 mark]***

You could have differentiated x with respect to y then used $\frac{dy}{dx} = \frac{1}{\left(\frac{dx}{dy}\right)}$ instead of using implicit differentiation here.

12 a) $\frac{dx}{d\theta} = \frac{\cos\theta}{2}$, $\frac{dy}{d\theta} = 2 \sin 2\theta$

So $\frac{dy}{dx} = \frac{dy}{d\theta} \div \frac{dx}{d\theta}$

$= 2 \sin 2\theta \div \frac{\cos\theta}{2} = \frac{4\sin 2\theta}{\cos\theta}$

[2 marks available — 1 mark for a correct method to find dy/dx, 1 mark for a correct expression for dy/dx]

b) When $\theta = \frac{\pi}{6}$, $\frac{dy}{dx} = \frac{4\sin\left(\frac{\pi}{3}\right)}{\cos\left(\frac{\pi}{6}\right)} = \frac{4\left(\frac{\sqrt{3}}{2}\right)}{\left(\frac{\sqrt{3}}{2}\right)} = 4$ ***[1 mark]***

$x = \frac{1}{2} \sin \frac{\pi}{6} - 3 = \frac{1}{2} \times \frac{1}{2} - 3 = \frac{1}{4} - 3 = -\frac{11}{4}$

$y = 5 - \cos \frac{\pi}{3} = 5 - \frac{1}{2} = \frac{9}{2}$

[1 mark for x and y values both correct]

So using $y - y_1 = m(x - x_1)$:

$y - \frac{9}{2} = 4\left(x + \frac{11}{4}\right) \Rightarrow 2y - 9 = 8x + 22$

So the equation in the correct form is $8x - 2y + 31 = 0$

[1 mark]

c) Substitute $x = \frac{\sin\theta}{2} - 3$ and $y = 5 - \cos 2\theta$ into $y = -8x - 20$:

$5 - \cos 2\theta = -8\left(\frac{\sin\theta}{2} - 3\right) - 20$ ***[1 mark]***

$\Rightarrow 1 - \cos 2\theta + 4\sin\theta = 0$

Using the double angle identity, $\cos 2\theta \equiv 1 - 2\sin^2\theta$,

so $1 - (1 - 2\sin^2\theta) + 4\sin\theta = 0$ ***[1 mark]***

$\Rightarrow 1 - 1 + 2\sin^2\theta + 4\sin\theta = 0$

$\Rightarrow 2\sin^2\theta + 4\sin\theta = 0$

$\Rightarrow \sin^2\theta + 2\sin\theta = 0$ ***[1 mark]***

$\Rightarrow \sin\theta(\sin\theta + 2) = 0$

This gives $\sin\theta = 0$ or $\sin\theta = -2$ ***[1 mark]***.

Since $\sin\theta$ must be between -1 and 1, $\sin\theta = -2$ is not valid.

So substitute $\sin\theta = 0$ into the expression for x:

$x = \frac{\sin\theta}{2} - 3 \Rightarrow x = 0 - 3 = -3$ ***[1 mark]***

Substitute $x = -3$ into $y = -8x - 20$

to get $y = -8(-3) - 20 = 4$ ***[1 mark]***

So P is the point $(-3, 4)$.

d) Rewrite the equation for y using the identity $\cos 2\theta \equiv 1 - 2\sin^2\theta$:

$y = 5 - \cos 2\theta = 5 - (1 - 2\sin^2\theta) = 4 + 2\sin^2\theta$ ***[1 mark]***

Now rearrange the equation for x to make $\sin\theta$ the subject:

$x = \frac{\sin\theta}{2} - 3 \Rightarrow 2x + 6 = \sin\theta$ ***[1 mark]***

Substitute this into the equation for y:

$y = 4 + 2\sin^2\theta = 4 + 2(2x + 6)^2$

$\Rightarrow y = 4 + 2(4x^2 + 24x + 36)$

$\Rightarrow y = 8x^2 + 48x + 76$ ***[1 mark]***

13 a) Use implicit differentiation to find $\frac{dy}{dx}$:

$\frac{d}{dx}x^3 + \frac{d}{dx}x^2y = \frac{d}{dx}y^2 - \frac{d}{dx}1$

$3x^2 + \frac{d}{dx}x^2y = \frac{d}{dx}y^2 - 0$

$3x^2 + \frac{d}{dx}x^2y = \frac{d}{dy}y^2\frac{dy}{dx}$

$3x^2 + \frac{d}{dx}x^2y = 2y\frac{dy}{dx}$

$3x^2 + x^2\frac{d}{dx}y + y\frac{d}{dx}x^2 = 2y\frac{dy}{dx}$

$3x^2 + x^2\frac{dy}{dx} + 2xy = 2y\frac{dy}{dx}$

Rearrange to make $\frac{dy}{dx}$ the subject:

$(2y - x^2)\frac{dy}{dx} = 3x^2 + 2xy$

$\frac{dy}{dx} = \frac{3x^2 + 2xy}{2y - x^2}$

[4 marks available — 1 mark for the correct differentiation of x^3 and –1, 1 mark for the correct differentiation of y^2, 1 mark for the correct differentiation of x^2y and 1 mark for rearranging to find the correct answer]

b) Substitute $x = 1$ into the original equation:

$x = 1 \Rightarrow (1)^3 + (1)^2y = y^2 - 1$ ***[1 mark]***

$\Rightarrow y^2 - y - 2 = 0$

$\Rightarrow (y - 2)(y + 1) = 0$

$\Rightarrow y = 2$ or $y = -1$

$a > b$, so $a = 2$, $b = -1$ ***[1 mark]***

c) At $Q\ (1, -1)$,

$\frac{dy}{dx} = \frac{3(1)^2 + 2(1)(-1)}{2(-1) - (1)^2} = \frac{3 - 2}{-2 - 1} = -\frac{1}{3}$ ***[1 mark]***

So the gradient of the normal at Q is $-1 \div -\frac{1}{3} = 3$ ***[1 mark]***

$y - y_1 = m(x - x_1)$

$\Rightarrow y + 1 = 3(x - 1)$

$\Rightarrow y = 3x - 4$ ***[1 mark]***

14 Use implicit differentiation to find $\frac{dy}{dx}$:

$\frac{d}{dx}(\sin\pi x) - \frac{d}{dx}\left(\cos\frac{\pi y}{2}\right) = \frac{d}{dx}(0.5)$

$\pi\cos\pi x - \frac{d}{dx}\left(\cos\frac{\pi y}{2}\right) = 0$ ***[1 mark]***

$\pi\cos\pi x - \frac{d}{dy}\left(\cos\frac{\pi y}{2}\right)\frac{dy}{dx} = 0$

$\pi\cos\pi x + \left(\frac{\pi}{2}\sin\frac{\pi y}{2}\right)\frac{dy}{dx} = 0$ ***[1 mark]***

Rearrange to make $\frac{dy}{dx}$ the subject:

$\frac{dy}{dx} = -\frac{\pi\cos\pi x}{\frac{\pi}{2}\sin\frac{\pi y}{2}} = -\frac{2\cos\pi x}{\sin\frac{\pi y}{2}}$

The stationary point is where the gradient is zero.

$\frac{dy}{dx} = 0 \Rightarrow -\frac{2\cos\pi x}{\sin\frac{\pi y}{2}} = 0$ ***[1 mark]*** $\Rightarrow \cos\pi x = 0$ ***[1 mark]***

$\Rightarrow x = \frac{1}{2}$ or $x = \frac{3}{2}$ in the range $0 \le x \le 2$ ***[1 mark]***

Put these values in the equation of the curve:

$x = \frac{3}{2} \Rightarrow \sin\frac{3\pi}{2} - \cos\frac{\pi y}{2} = 0.5$

$\Rightarrow -1 - \cos\frac{\pi y}{2} = 0.5$

$\Rightarrow \cos\frac{\pi y}{2} = -1.5$

So y has no solutions when $x = \frac{3}{2}$ ***[1 mark]***

$x = \frac{1}{2} \Rightarrow \sin\frac{\pi}{2} - \cos\frac{\pi y}{2} = 0.5$

$\Rightarrow 1 - \cos\frac{\pi y}{2} = 0.5$

$\Rightarrow \cos\frac{\pi y}{2} = 0.5$

$\Rightarrow \frac{\pi y}{2} = \frac{\pi}{3}$

$\Rightarrow y = \frac{2}{3}$ in the range $0 \le y \le 2$ ***[1 mark]***

So the stationary point of the graph of $\sin\pi x - \cos\frac{\pi y}{2} = 0.5$ for the given ranges of x and y is at $\left(\frac{1}{2}, \frac{2}{3}\right)$.

Pages 51-55: Integration 1

1 $\int\left(\frac{x^2 + 3}{\sqrt{x}}\right)dx = \int\left(x^{\frac{3}{2}} + 3x^{-\frac{1}{2}}\right)dx = \frac{x^{\frac{5}{2}}}{(5/2)} + \frac{3x^{\frac{1}{2}}}{(1/2)} + C$

$= \frac{2}{5}x^{\frac{5}{2}} + 6x^{\frac{1}{2}} + C = \frac{2}{5}\sqrt{x^5} + 6\sqrt{x} + C$

[3 marks available — 1 mark for writing both terms as powers of x, 1 mark for increasing the power of one term by 1, 1 mark for the correct integrated terms and adding C]

2 To find f(x), integrate f'(x):

$f(x) = \int\left(2x + 5\sqrt{x} + \frac{6}{x^2}\right)dx = \int\left(2x + 5x^{\frac{1}{2}} + 6x^{-2}\right)dx$

$= \frac{2x^2}{2} + 5\left(\frac{x^{\frac{3}{2}}}{\left(\frac{3}{2}\right)}\right) + \left(\frac{6x^{-1}}{-1}\right) + C$

$f(x) = x^2 + \frac{10\sqrt{x^3}}{3} - \frac{6}{x} + C$

[4 marks available for the above working — 1 mark for writing all terms as powers of x, 1 mark for increasing the power of one term by 1, 1 mark for two correct simplified terms, 1 mark for the third correct integrated term and adding C]

You've been given a point on the curve so calculate the value of C:

If $y = 7$ when $x = 3$, then

$3^2 + \frac{10\sqrt{3^3}}{3} - \frac{6}{3} + C = 7$ ***[1 mark]***

$9 + 10\sqrt{3} - 2 + C = 7$

$7 + 10\sqrt{3} + C = 7 \Rightarrow C = -10\sqrt{3}$

$f(x) = x^2 + \frac{10\sqrt{x^3}}{3} - \frac{6}{x} - 10\sqrt{3}$ ***[1 mark]***

3 To find the area of region A, you need to integrate the function between $x = 2$ and $x = 4$:

Area $= \int_2^4 \frac{2}{\sqrt{x^3}}dx = \int_2^4 2x^{-\frac{3}{2}}dx = \left[-2(2x^{-\frac{1}{2}})\right]_2^4 = \left[\frac{-4}{x^{\frac{1}{2}}}\right]_2^4 = \left[\frac{-4}{\sqrt{x}}\right]_2^4$

$= \left(\frac{-4}{\sqrt{4}}\right) - \left(\frac{-4}{\sqrt{2}}\right) = \frac{-4}{2} + \frac{4}{\sqrt{2}} = -2 + \frac{4\sqrt{2}}{2}$

$= 2\sqrt{2} - 2$ as required

[5 marks available — 1 mark for writing down the correct integral to find, 1 mark for integrating correctly, 1 mark for correct handling of the limits, 1 mark for rationalising the denominator, 1 mark for rearranging to give the answer in the correct form]

4 $\int_p^{4p}\left(\frac{1}{\sqrt{x}}-4x^3\right)dx=\int_p^{4p}(x^{-\frac{1}{2}}-4x^3)\,dx=\left[\frac{x^{\frac{1}{2}}}{\frac{1}{2}}-\frac{4x^4}{4}\right]_p^{4p}$

$=\left[2x^{\frac{1}{2}}-x^4\right]_p^{4p}=\left[2\sqrt{x}-x^4\right]_p^{4p}$

$=(2\sqrt{4p}-(4p)^4)-(2\sqrt{p}-p^4)$

$=(4\sqrt{p}-256p^4)-(2\sqrt{p}-p^4)$

$=2\sqrt{p}-255p^4$

[4 marks available — 1 mark for increasing the power of one term by 1, 1 mark for the correct integrated terms, 1 mark for correct handling of the limits, 1 mark for simplifying to get the final answer]

5 Find the value of the integral in terms of p:

$\int_{2p}^{6p}\frac{x^3+4x^2}{x^3}\,dx=\int_{2p}^{6p}\left(1+\frac{4}{x}\right)dx=[x+4\ln x]_{2p}^{6p}$

$=(6p+4\ln 6p)-(2p+4\ln 2p)$

$=4p+4\ln\frac{6p}{2p}=4p+4\ln 3$

You know the value of the integral, so set these expressions equal to each other and solve for p:

$4p+4\ln 3=4\ln 12$

$p=\ln 12-\ln 3=\ln\frac{12}{3}=\ln 4$

[4 marks available — 1 mark for simplifying the fraction and integrating, 1 mark for the correct integral, 1 mark for substituting in the limits correctly and simplifying, 1 mark for setting the expressions equal to each other and solving to find p]

6 First, rewrite as partial fractions:

$\frac{1}{x(3x-2)}\equiv\frac{A}{x}+\frac{B}{3x-2}\Rightarrow 1\equiv A(3x-2)+Bx$ ***[1 mark]***

Substituting $x=0$ gives $A=-\frac{1}{2}$

Substituting $x=\frac{2}{3}$ gives $B=\frac{3}{2}$ ***[1 mark for both]***

$\Rightarrow\frac{1}{x(3x-2)}\equiv\frac{3}{2(3x-2)}-\frac{1}{2x}$

So $\int\frac{1}{x(3x-2)}\,dx=\int\left(\frac{3}{2(3x-2)}-\frac{1}{2x}\right)dx$ ***[1 mark]***

$=\int\left(\frac{3}{2}(3x-2)^{-1}-\frac{1}{2}x^{-1}\right)dx$

$=\frac{1}{2}\ln|3x-2|-\frac{1}{2}\ln|x|+C$

$=\frac{1}{2}\ln\left|\frac{3x-2}{x}\right|+C$

[2 marks for the correct answer — deduct 1 mark if the constant is missing]

You could also write C as ln k, making your final answer $\frac{1}{2}\ln\left|\frac{k(3x-2)}{x}\right|$. As long as you include the constant, then you'll get all the marks.

7 a) $\frac{3x+5}{(3x+1)(1-x)}\equiv\frac{A}{3x+1}+\frac{B}{1-x}$ ***[1 mark]***

$\Rightarrow 3x+5\equiv A(1-x)+B(3x+1)$

Substituting $x=1$ gives $B=2$

Substituting $x=-\frac{1}{3}$ gives $A=3$ ***[1 mark for both]***

$\Rightarrow\frac{3x+5}{(3x+1)(1-x)}\equiv\frac{3}{(3x+1)}+\frac{2}{(1-x)}$ ***[1 mark]***

b) $\int_{-2}^{4}\frac{3x+5}{(3x+1)(1-x)}\,dx=\int_{-2}^{4}\left(\frac{3}{(3x+1)}+\frac{2}{(1-x)}\right)dx$

$=[\ln|3x+1|-2\ln|1-x|]_{-2}^{4}$

$=(\ln 13-2\ln|-3|)-(\ln|-5|-2\ln 3)$

$=\ln 13-2\ln 3-\ln 5+2\ln 3$

$=\ln 13-\ln 5=\ln\frac{13}{5}$

[3 marks available — 1 mark for the correct integral, 1 mark for substituting in the limits, 1 mark for simplifying to give the answer in the correct form]

8 $4x^2+4x-3=(2x+3)(2x-1)$, so $\frac{4x-10}{4x^2+4x-3}\equiv\frac{4x-10}{(2x+3)(2x-1)}$

Rewrite this as partial fractions:

$\frac{4x-10}{(2x+3)(2x-1)}\equiv\frac{A}{2x+3}+\frac{B}{2x-1}$

$\Rightarrow 4x-10\equiv A(2x-1)+B(2x+3)$

Substituting $x=\frac{1}{2}$ gives $B=-2$

Substituting $x=-\frac{3}{2}$ gives $A=4$

$\Rightarrow\frac{4x-10}{(2x+3)(2x-1)}\equiv\frac{4}{(2x+3)}-\frac{2}{(2x-1)}$

Now integrate:

$\int_{-1}^{1}\frac{4x-10}{4x^2+4x-3}\,dx=\int_{-1}^{1}\left(\frac{4}{(2x+3)}-\frac{2}{(2x-1)}\right)dx$

$=[2\ln|2x+3|-\ln|2x-1|]_{-1}^{1}$

$=(2\ln 5-\ln 1)-(2\ln 1-\ln|-3|)$

$=2\ln 5+\ln 3=\ln(5^2\times 3)=\ln 75$

So k = 75.

[7 marks available — 1 mark for factorising the denominator, 1 mark for writing as partial fractions with the correct denominators, 1 mark for finding A or B, 1 mark for the correct partial fractions, 1 mark for the correct integral, 1 mark for substituting in the limits, 1 mark for simplifying to give the answer in the correct form]

9 $\int_{\frac{\pi}{12}}^{\frac{\pi}{8}}\sin 2x\,dx=\left[-\frac{1}{2}\cos 2x\right]_{\frac{\pi}{12}}^{\frac{\pi}{8}}$

$=\left(-\frac{1}{2}\cos\left(\frac{\pi}{4}\right)\right)-\left(-\frac{1}{2}\cos\left(\frac{\pi}{6}\right)\right)$

$=-\frac{1}{2\sqrt{2}}+\frac{\sqrt{3}}{4}=\frac{\sqrt{3}-\sqrt{2}}{4}$

[3 marks available — 1 mark for integrating correctly, 1 mark for substituting in the limits, 1 mark for the correct answer in surd form]

10 Use the identity $\sec^2 x\equiv 1+\tan^2 x$ to write $3\tan^2\frac{x}{2}+3$ as $3\sec^2\frac{x}{2}$.

The integral becomes:

$\int\left(3\tan^2\left(\frac{x}{2}\right)+3\right)dx=\int 3\sec^2\left(\frac{x}{2}\right)dx=6\tan\left(\frac{x}{2}\right)+C$

[3 marks available — 1 mark for rewriting the integral in terms of $\sec^2\frac{x}{2}$, 1 mark for integrating correctly, 1 mark for the correct answer including + C]

11 $\frac{d}{dx}(e^{\tan x})=\sec^2 x\,e^{\tan x}$, so the integral is of the form

$\int\frac{du}{dx}f'(u)\,dx=f(u)+C$, so $\int\sec^2 x\,e^{\tan x}\,dx=e^{\tan x}+C$

[2 marks available — 1 mark for integrating correctly, 1 mark for the correct answer including + C]

12 Integrate the curve $y=\frac{2}{3(\sqrt[3]{5x-2})}$ with respect to x between 2 and 5.8 to find the shaded region:

$\int_2^{5.8}\frac{2}{3(\sqrt[3]{5x-2})}\,dx=\int_2^{5.8}\frac{2}{3}(5x-2)^{-\frac{1}{3}}\,dx$ ***[1 mark]***

$=\left[\frac{1}{5}\cdot\frac{2}{3}\cdot\frac{3}{2}(5x-2)^{\frac{2}{3}}\right]_2^{5.8}$ ***[1 mark]***

$=\left[\frac{1}{5}(\sqrt[3]{5x-2})^2\right]_2^{5.8}$

$=\frac{1}{5}(\sqrt[3]{27})^2-\frac{1}{5}(\sqrt[3]{8})^2$ ***[1 mark]***

$=\frac{3^2}{5}-\frac{2^2}{5}=\frac{5}{5}=1$ ***[1 mark]***

13 Evaluate the integral, treating k as a constant:

$\int_{\sqrt{2}}^{2}(8x^3-2kx)\,dx=\left[\frac{8x^4}{4}-\frac{2kx^2}{2}\right]_{\sqrt{2}}^{2}=[2x^4-kx^2]_{\sqrt{2}}^{2}$

$=(2(2)^4-k(2)^2)-(2(\sqrt{2})^4-k(\sqrt{2})^2)$

$=(32-4k)-(8-2k)=24-2k$

You know that the value of this integral is $2k^2$, so set this expression equal to $2k^2$ and solve to find k:

$24-2k=2k^2$

$0=2k^2+2k-24\Rightarrow k^2+k-12=0\Rightarrow(k+4)(k-3)=0$

So $k=-4$ or $k=3$

[5 marks available — 1 mark for increasing the power of one term by 1, 1 mark for the correct integrated terms, 1 mark for substituting in the limits, 1 mark for setting this expression equal to $2k^2$, 1 mark for solving the quadratic to find both values of k]

14 To find the shaded area, you need to integrate the function between –1 and 0.5 and add it to the integral of the function between 0.5 and 2 (making this value positive first).

$$\int_{-1}^{0.5}(2x^3-3x^2-11x+6)\,dx=\left[\frac{2x^4}{4}-\frac{3x^3}{3}-\frac{11x^2}{2}+6x\right]_{-1}^{0.5}$$
$$=\left[\frac{x^4}{2}-x^3-\frac{11}{2}x^2+6x\right]_{-1}^{0.5}$$
$$=\left(\frac{(0.5)^4}{2}-(0.5)^3-\frac{11}{2}(0.5)^2+6(0.5)\right)-\left(\frac{(-1)^4}{2}-(-1)^3-\frac{11}{2}(-1)^2+6(-1)\right)$$
$$=1.53125-(-10)=11.53125$$

So the area between –1 and 0.5 is 11.53125.

$$\int_{0.5}^{2}(2x^3-3x^2-11x+6)\,dx=\left[\frac{x^4}{2}-x^3-\frac{11}{2}x^2+6x\right]_{0.5}^{2}$$
$$=\left(\frac{(2)^4}{2}-(2)^3-\frac{11}{2}(2)^2+6(2)\right)-1.53125$$
$$=-10-1.53125=-11.53125$$

So the area between 0.5 and 2 is 11.53125.
So area = 11.53125 + 11.53125 = 23.0625

[6 marks available — 1 mark for considering the area above and below the x-axes separately, 1 mark for increasing the power of one term by 1, 1 mark for the correct integral, 1 mark for finding the area between –1 and 0.5, 1 mark for finding the area between 0.5 and 2, 1 mark for adding the areas to get the correct answer]

If you'd just integrated between –1 and 2, you'd have ended up with an answer of 0, as the areas cancel each other out.

15 a) A and B are the points where the two lines intersect, so

$\frac{8}{x^2}=9-x^2$ ***[1 mark]*** $\Rightarrow 8=9x^2-x^4$

$\Rightarrow x^4-9x^2+8=0 \Rightarrow (x^2-8)(x^2-1)=0$ ***[1 mark]***

$x^2=8 \Rightarrow x=\pm\sqrt{8}=\pm2\sqrt{2}=2\sqrt{2}$ as $x\geq0$

$x^2=1 \Rightarrow x=\pm1=1$ as $x\geq0$ ***[1 mark for both values of x]***

When $x=1$, $y=8$, and when $x=2\sqrt{2}$, $y=1$

So $A=(1, 8)$ ***[1 mark]*** and $B=(2\sqrt{2}, 1)$ ***[1 mark]***.

b) The shaded region is the area under $y=9-x^2$ minus the area under $y=\frac{8}{x^2}$ from $x=1$ to $x=2\sqrt{2}$, so integrate:

$$\int_1^{2\sqrt{2}}\left(9-x^2-\frac{8}{x^2}\right)dx=\left[9x-\frac{x^3}{3}+\frac{8}{x}\right]_1^{2\sqrt{2}}$$
$$=\left(18\sqrt{2}-\frac{(2\sqrt{2})^3}{3}+\frac{8}{2\sqrt{2}}\right)-\left(9-\frac{1}{3}+8\right)$$
$$=18\sqrt{2}-\frac{16\sqrt{2}}{3}+2\sqrt{2}-9+\frac{1}{3}-8$$
$$=\frac{44}{3}\sqrt{2}-\frac{50}{3}$$

[4 marks available — 1 mark for increasing the power of one term by 1, 1 mark for the correct integral, 1 mark for correct handling of the limits, 1 mark for simplifying to get the final answer]

16 First find the equation of line N — it's a normal to the curve, so differentiate: $\frac{dy}{dx}=\frac{1}{2}x^{-\frac{1}{2}}-x$

[1 mark for differentiating, 1 mark for correct derivative]

When $x=1$, $\frac{dy}{dx}=\frac{1}{2}-1=-\frac{1}{2}$ ***[1 mark]***, so the gradient of the normal to the curve at this point is $-1\div-\frac{1}{2}=2$ ***[1 mark]***.

So the equation of the normal is:

$y-\frac{3}{2}=2(x-1)\Rightarrow y=2x-\frac{1}{2}$ ***[1 mark]***.

So the area you need is:

$$\int_0^1\left[\left(\sqrt{x}-\frac{1}{2}x^2+1\right)-\left(2x-\frac{1}{2}\right)\right]dx \quad \textit{[1 mark]}$$
$$=\int_0^1\left(x^{\frac{1}{2}}-\frac{1}{2}x^2-2x+\frac{3}{2}\right)dx$$
$$=\left[\frac{2}{3}x^{\frac{3}{2}}-\frac{1}{6}x^3-x^2+\frac{3}{2}x\right]_0^1 \quad \textit{[1 mark]}$$
$$=\left(\frac{2}{3}-\frac{1}{6}-1+\frac{3}{2}\right)-0=1$$

[1 mark for substituting in the limits correctly, 1 mark for the correct answer]

You could have done this one by working out each bit separately

17 a) The graph of $y=x^2+1$ crosses the y-axis at $y=0+1=1$. So the area you need to find is shown below:

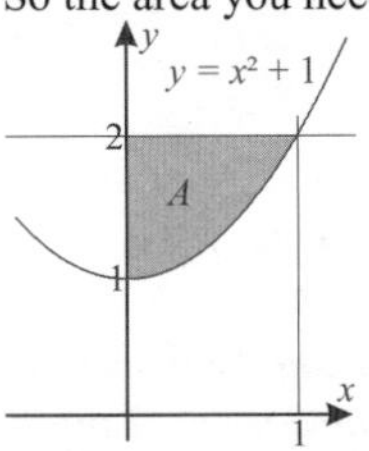

This is the integral $\int_1^2 x\,dy$ (the limits show that y goes from 1 to 2).

Rearrange $y=x^2+1$ to find x in terms of y: $x=\sqrt{y-1}$

So the required area is given by $A=\int_1^2\sqrt{y-1}\,dy$.

[2 marks available — 1 mark for finding x in terms of y, and 1 mark showing how to find the limits]

b) Integrating with respect to y gives:

$$A=\int_1^2\sqrt{y-1}\,dy$$
$$=\int_1^2(y-1)^{\frac{1}{2}}dy$$
$$=\left[\frac{2}{3}(y-1)^{\frac{3}{2}}\right]_1^2=\frac{2}{3}-0=\frac{2}{3}$$

[3 marks available — 1 mark for integrating correctly, 1 mark for substituting in the limits, 1 mark for the correct answer]

Pages 56-59: Integration 2

1 If $x=\sin\theta$, then $\frac{dx}{d\theta}=\cos\theta$, so $dx=d\theta\cos\theta$ ***[1 mark]***.

Change the limits: as $x=\sin\theta$, $\theta=\sin^{-1}x$,

so when $x=0$, $\theta=0$ and when $x=\frac{1}{2}$, $\theta=\frac{\pi}{6}$ ***[1 mark]***.

Putting all this into the integral gives:

$$\int_0^{\frac{1}{2}}\frac{x}{1-x^2}dx=\int_0^{\frac{\pi}{6}}\frac{\sin\theta}{1-\sin^2\theta}\cos\theta\,d\theta \quad \textit{[1 mark]}$$

Using the identity $\sin^2\theta+\cos^2\theta\equiv1$, replace $1-\sin^2\theta$:

$$\int_0^{\frac{\pi}{6}}\frac{\sin\theta\cos\theta}{\cos^2\theta}d\theta=\int_0^{\frac{\pi}{6}}\frac{\sin\theta}{\cos\theta}d\theta=\int_0^{\frac{\pi}{6}}\tan\theta\,d\theta \quad \textit{[1 mark]}$$
$$=[-\ln|\cos\theta|]_0^{\frac{\pi}{6}} \quad \textit{[1 mark]}$$
$$=-\ln\left|\cos\frac{\pi}{6}\right|+\ln|\cos0| \quad \textit{[1 mark]}$$
$$=-\ln\frac{\sqrt{3}}{2}+\ln1=-\ln\sqrt{3}+\ln2$$
$$=\ln2-\ln\sqrt{3}\left(=\ln\frac{2}{\sqrt{3}}\right) \quad \textit{[1 mark]}$$

Be careful when the substitution is of the form x = f(θ) rather than θ = f(x) — when you change the limits, you need to find the inverse of f(θ) then put in the given values of x.

2 As $u=\ln x$, $\frac{du}{dx}=\frac{1}{x}$, so $x\,du=dx$ ***[1 mark]***. The limits $x=1$ and $x=2$ become $u=\ln1=0$ and $u=\ln2$ ***[1 mark]***.

$\left(\frac{\ln x}{\sqrt{x}}\right)^2=\frac{(\ln x)^2}{x}$. So the integral is:

$$\int_0^{\ln2}\frac{u^2}{x}x\,du=\int_0^{\ln2}u^2\,du \quad \textit{[1 mark]}$$
$$=\left[\frac{u^3}{3}\right]_0^{\ln2} \quad \textit{[1 mark]}$$
$$=\frac{(\ln2)^3}{3}=0.111\ (3\text{ s.f.}) \quad \textit{[1 mark]}$$

3 Let $u=4x$, so $\frac{du}{dx}=4$. Let $\frac{dv}{dx}=e^{-2x}$, so $v=-\frac{1}{2}e^{-2x}$.

Putting this into the formula for integration by parts gives:

$$\int4xe^{-2x}dx=\left[4x\left(-\frac{1}{2}e^{-2x}\right)\right]-\int4\left(-\frac{1}{2}e^{-2x}\right)dx$$
$$=-2xe^{-2x}+\int2e^{-2x}dx$$
$$=-2xe^{-2x}-e^{-2x}+C\ (=-e^{-2x}(2x+1)+C)$$

[4 marks available — 1 mark for correct choice of u and dv/dx, 1 mark for correct differentiation and integration to obtain du/dx and v, 1 mark for correct integration by parts method, 1 mark for answer including + C]

4 a) Let $u = x$, so $\frac{du}{dx} = 1$.
Let $\frac{dv}{dx} = \sin 4x$, so $v = -\frac{1}{4}\cos 4x$
Using integration by parts,
$$\int x\sin 4x\, dx = -\frac{1}{4}x\cos 4x - \int -\frac{1}{4}\cos 4x\, dx$$
$$= -\frac{1}{4}x\cos 4x + \frac{1}{16}\sin 4x + C$$
[4 marks available — 1 mark for correct choice of u and dv/dx, 1 mark for correct differentiation and integration to obtain du/dx and v, 1 mark for correct integration by parts method, 1 mark for correct answer including + C]

b) Let $u = x^2$, so $\frac{du}{dx} = 2x$.
Let $\frac{dv}{dx} = \cos 4x$, so $v = \frac{1}{4}\sin 4x$
Using integration by parts,
$$\int x^2\cos 4x\, dx = \frac{1}{4}x^2\sin 4x - \int \frac{1}{2}x\sin 4x\, dx$$
$$= \frac{1}{4}x^2\sin 4x - \frac{1}{2}\left(-\frac{1}{4}x\cos 4x + \frac{1}{16}\sin 4x\right) + C$$
$$= \frac{1}{4}x^2\sin 4x + \frac{1}{8}x\cos 4x - \frac{1}{32}\sin 4x + C$$
[4 marks available — 1 mark for correct choice of u and dv/dx, 1 mark for correct differentiation and integration to obtain du/dx and v, 1 mark for correct integration by parts method, 1 mark for correct answer including + C]

You've already worked out $\int x\sin 4x\, dx$ in part a), so you can use this answer in part b).

5 Let $u = \ln x$, so $\frac{du}{dx} = \frac{1}{x}$.
Let $\frac{dv}{dx} = \frac{1}{2x^2} = \frac{1}{2}x^{-2}$, so $v = -\frac{1}{2}x^{-1}$
Using integration by parts,
$$\int_1^4 \frac{\ln x}{2x^2}\, dx = \left[-\frac{\ln x}{2x}\right]_1^4 - \int_1^4 -\frac{1}{2x^2}\, dx = \left[-\frac{\ln x}{2x}\right]_1^4 - \left[\frac{1}{2x}\right]_1^4$$
$$= \left[-\frac{\ln 4}{8} - -\frac{\ln 1}{2}\right] - \left[\frac{1}{8} - \frac{1}{2}\right] = \frac{3}{8} - \frac{\ln 4}{8} \quad \left(= \frac{3 - \ln 4}{8}\right)$$
[6 marks available — 1 mark for correct choice of u and dv/dx, 1 mark for correct differentiation and integration to obtain du/dx and v, 1 mark for correct integration by parts method, 1 mark for correct integral, 1 mark for substituting in the limits, 1 mark for answer]

6 a) $\frac{dN}{dt} = k\sqrt{N}$, $k > 0$ ***[1 mark for LHS, 1 mark for RHS]***
When $N = 36$, $\frac{dN}{dt} = 0.36$. Putting these values into the equation gives $0.36 = k\sqrt{36}$ ***[1 mark]*** $= 6k \Rightarrow k = 0.06$ (the population is increasing so ignore the negative square root).
So the differential equation is $\frac{dN}{dt} = 0.06\sqrt{N}$ ***[1 mark]***.

b) (i) $\frac{dN}{dt} = \frac{kN}{\sqrt{t}} \Rightarrow \int \frac{1}{N}\, dN = \int \frac{k}{\sqrt{t}}\, dt$
$\ln|N| = 2k\sqrt{t} + C$ ***[1 mark]***
$\Rightarrow N = e^{2k\sqrt{t} + C} = Ae^{2k\sqrt{t}}$, where $A = e^C$ ***[1 mark]***
For the initial population, $t = 0$, so $N = 25$ when $t = 0$.
Putting these values into the equation: $25 = Ae^0 \Rightarrow 25 = A$, so the equation for N is: $N = 25e^{2k\sqrt{t}}$ ***[1 mark]***.

(ii) When initial population has doubled, $N = 50$ ***[1 mark]***.
Put this value and the value for k into the equation and solve for t:
$50 = 25e^{2(0.05)\sqrt{t}} \Rightarrow 2 = e^{0.1\sqrt{t}} \Rightarrow \ln 2 = 0.1\sqrt{t}$ ***[1 mark]***
$10\ln 2 = \sqrt{t} \Rightarrow (10\ln 2)^2 = t \Rightarrow t = 48.045$
So it will take 48 weeks ***[1 mark]*** (to the nearest week) for the population to double.

7 a) First solve the differential equation to find S:
$$\frac{dS}{dt} = k\sqrt{S} \Rightarrow \frac{1}{\sqrt{S}}\, dS = k\, dt$$
$$\Rightarrow \int S^{-\frac{1}{2}}\, dS = \int k\, dt$$
$\Rightarrow 2S^{\frac{1}{2}} = kt + C$ ***[1 mark]***
$\Rightarrow S = \left(\frac{1}{2}(kt + C)\right)^2 = \frac{1}{4}(kt + C)^2$ ***[1 mark]***
At the start of the campaign, $t = 0$.
Putting $t = 0$ and $S = 81$ into the equation gives:
$81 = \frac{1}{4}(0 + C)^2 \Rightarrow 324 = C^2 \Rightarrow C = 18$ (C must be positive, otherwise the sales would be decreasing).
This gives the equation $S = \frac{1}{4}(kt + 18)^2$ ***[1 mark]***.

b) When $t = 0$, $S = 81$ and $\frac{dS}{dt} = 18$.
Substituting this into $\frac{dS}{dt} = k\sqrt{S}$ gives $k = 2$.
Using $S = \frac{1}{4}(kt + 18)^2$ with $t = 5$ and $k = 2$ gives $\frac{1}{4}((5 \times 2) + 18)^2 = 196$ kg sold.
[3 marks available — 1 mark for finding the value of k, 1 mark for substituting correct values of t and k, 1 mark for answer]

c) To find the value of t when $S = 225$, solve the equation $225 = \frac{1}{4}(2t + 18)^2$ ***[1 mark]***:
$225 = \frac{1}{4}(2t + 18)^2 \Rightarrow 900 = (2t + 18)^2$
$\Rightarrow 30 = 2t + 18 \Rightarrow 12 = 2t \Rightarrow 6 = t$
So it will be 6 days ***[1 mark]*** before 225 kg of cheese is sold.

8 a) (i) $\frac{dr}{dt} = -krt$, $k > 0$ ***[1 mark for RHS, 1 mark for LHS]***
(ii) S = area of curved surface + area of circular surface
$= \frac{1}{2}(4\pi r^2) + \pi r^2 = 3\pi r^2$ ***[1 mark]***
So $\frac{dS}{dr} = 6\pi r$ ***[1 mark]***
$\frac{dS}{dt} = \frac{dS}{dr} \times \frac{dr}{dt}$ ***[1 mark]***
$= 6\pi r \times -krt = -6\pi kr^2 t = -2kt(3\pi r^2) = -2ktS$
[1 mark for correct substitution and simplification to required answer]

b) (i) $\frac{dS}{dt} = -2ktS \Rightarrow \frac{dS}{S} = -2kt\, dt$
$\Rightarrow \int \frac{1}{S}\, dS = \int -2kt\, dt$
$\Rightarrow \ln S = -kt^2 + \ln A$
[1 mark for correct integration of both sides, plus a constant term]
$\Rightarrow S = e^{-kt^2 + \ln A} = Ae^{-kt^2}$ ***[1 mark]***
$S = 200$ at $t = 10 \Rightarrow 200 = Ae^{-100k}$
$S = 50$ at $t = 30 \Rightarrow 50 = Ae^{-900k}$
$\Rightarrow Ae^{-100k} = 4Ae^{-900k}$ ***[1 mark]***
$\Rightarrow e^{-100k} = 4e^{-900k}$
$\Rightarrow -100k = \ln 4 - 900k$
$\Rightarrow 800k = \ln 4$
$\Rightarrow k = 0.00173$ (3 s.f.) ***[1 mark]***
So $200 = Ae^{-100k} = Ae^{-0.173} = 0.841A$
$\Rightarrow A = 238$ (3 s.f.) ***[1 mark]***
So $S = 238e^{-0.00173t^2}$

(ii) The initial surface area is given when $t = 0$
$\Rightarrow S = 238e^0 = 238$ cm^2 (3 s.f.) ***[1 mark]***

c) E.g. The differential equation for the hemisphere was calculated using an expression for its surface area.
The expression for the surface area of a full sphere will be different to the expression for a hemisphere of the same radius, so this differential equation will not be appropriate ***[1 mark for a sensible comment]***.

Pages 60-64: Numerical Methods

1 a) To find the inverse, let $y = f(x)$, so $y = 4(x^2 - 1)$.
Now make x the subject:

$y = 4(x^2 - 1) \Rightarrow \frac{y}{4} = x^2 - 1 \Rightarrow \frac{y}{4} + 1 = x^2$

So $x = \sqrt{\frac{y}{4} + 1}$ ***[1 mark]*** (you can ignore the negative square root, as the domain of $f(x)$ is $x \geq 0$).

Finally, replace y with x and x with $f^{-1}(x)$:

$f^{-1}(x) = \sqrt{\frac{x}{4} + 1}$ ***[1 mark for correct inverse]***.

$y = f^{-1}(x)$ is a reflection of $y = f(x)$ in the line $y = x$ ***[1 mark]***, so the point at which the lines $y = f(x)$ and $y = f^{-1}(x)$ meet is also the point where $y = f^{-1}(x)$ meets the line $y = x$.

At this point, $x = \sqrt{\frac{x}{4} + 1}$, so $\sqrt{\frac{x}{4} + 1} - x = 0$

[1 mark for setting $f^{-1}(x)$ equal to x and rearranging].

b) Let $g(x) = \sqrt{\frac{x}{4} + 1} - x$

If there is a root in the interval $1 < x < 2$ then there will be a change of sign for $g(x)$ between 1 and 2:

$g(1) = \sqrt{\frac{1}{4} + 1} - 1 = 0.1180...$

$g(2) = \sqrt{\frac{2}{4} + 1} - 2 = -0.7752...$ ***[1 mark for both]***

There is a change of sign and the function is continuous over this interval, so there is a root in the interval $1 < x < 2$ ***[1 mark]***.

c) $x_{n+1} = \sqrt{\frac{x_n}{4} + 1}$, and $x_0 = 1$, so:

$x_1 = \sqrt{\frac{1}{4} + 1} = 1.1180...$ ***[1 mark]***

$x_2 = \sqrt{\frac{1.1180...}{4} + 1} = 1.1311...$

$x_3 = \sqrt{\frac{1.1311...}{4} + 1} = 1.1326...$ ***[1 mark]***

So $x = 1.13$ to 3 s.f. ***[1 mark]***.

d) No. If you sketch the line $y = x$ on the graph, you can see that it does not cross the curve $y = f^{-1}(x)$ more than once, so there is only one root of the equation $\sqrt{\frac{x}{4} + 1} - x = 0$.

[1 mark for 'No' with suitable explanation]

2 a) When the curve and line intersect, $6^x = x + 2 \Rightarrow 6^x - x - 2 = 0$.
Let $f(x) = 6^x - x - 2$. If there is a root in the interval $[0.5, 1]$ then there will be a change of sign for $f(x)$ between 0.5 and 1:

$f(0.5) = 6^{0.5} - 0.5 - 2 = -0.0505...$

$f(1) = 6^1 - 1 - 2 = 3$ ***[1 mark for both]***

There is a change of sign and the function is continuous over this interval, so there is a root in the interval $[0.5, 1]$ ***[1 mark]***.

b) Substitute $f(x) = 6^x - x - 2$ into the Newton-Raphson formula:

$$x_{n+1} = x_n - \frac{f(x_n)}{f'(x_n)} = x_n - \frac{6^{x_n} - x_n - 2}{6^{x_n} \ln 6 - 1}$$

$$= \frac{x_n(6^{x_n} \ln 6 - 1)}{6^{x_n} \ln 6 - 1} - \frac{6^{x_n} - x_n - 2}{6^{x_n} \ln 6 - 1}$$

$$= \frac{x_n 6^{x_n} \ln 6 - x_n - 6^{x_n} + x_n + 2}{6^{x_n} \ln 6 - 1}$$

$$= \frac{6^{x_n}(x_n \ln 6 - 1) + 2}{6^{x_n} \ln 6 - 1} \text{ as required}$$

[4 marks available — 1 mark for differentiating f(x) correctly, 1 mark for substituting everything into the Newton-Raphson formula, 1 mark for putting x_n and $f(x_n)$ over a common denominator, 1 mark for factorising the numerator]

c) $x_0 = 0.5$

$x_1 = \frac{6^{(0.5)}((0.5)\ln 6 - 1) + 2}{6^{(0.5)} \ln 6 - 1} = 0.514904...$

$x_2 = 0.514653...$

$x_3 = 0.514653...$

x_2 and x_3 are both the same to at least 5 s.f. so the x-coordinate of P is 0.5147 (4 s.f.)

[3 marks available — 1 mark for putting $x_0 = 0.5$ into the Newton-Raphson formula from part b), 1 mark for repeated iterations until all values round to the same number to an appropriate number of significant figures, 1 mark for the correct answer]

If you put 0.5 into your calculator and press =, then input (6^{ANS} × (ANS × ln 6 − 1) + 2) ÷ (6^{ANS} × ln 6 − 1) and keep pressing =, you'll get the iterative sequence without having to type it in each time.

d) From above, $x = 0.5147$ to 4 s.f.
If $x = 0.5147$ to 4 s.f., the upper and lower bounds are 0.51475 and 0.51465 ***[1 mark]*** — any value in this range would be rounded to 0.5147. $f(x) = 6^x - x - 2$, and at point P, $f(x) = 0$.

$f(0.51475) = 0.000338...$ and $f(0.51465) = -0.0000116...$

[1 mark for both f(0.51475) positive and f(0.51465) negative].

There is a change of sign, and since $f(x)$ is continuous there must be a root in this interval ***[1 mark]***.

e) E.g. If the tangent has a gradient of 0, the denominator of the fraction in the iteration formula will be 0 so the Newton-Raphson method will fail for this starting value as no value of x_1 can be found. / If the tangent has a gradient of 0, it is a horizontal line so will never cross the x-axis so the Newton-Raphson method will fail. ***[1 mark for any suitable explanation]***

3 a) Substitute the values you're given for x_0, $f(x_0)$ and $f'(x_0)$ into the Newton-Raphson formula:

$x_1 = x_0 - \frac{f(x_0)}{f'(x_0)} = -1.5 - \frac{-1.625}{9.75}$

$= -1.33333... = -1.333$ (4 s.f.)

[2 marks available — 1 mark for substituting the values into the Newton-Raphson formula correctly, 1 mark for the correct answer]

b) The denominator of the fraction in the Newton-Raphson formula is $f'(x) = 3x^2 - 2x$ ***[1 mark]***.

When $x = \frac{2}{3}$, $3x^2 - 2x = 3(\frac{2}{3})^2 - 2(\frac{2}{3}) = 0$

— so the denominator is 0, which means the Newton-Raphson method fails as no value of x_1 can be found ***[1 mark]***.

4 a) The trapezium rule is given by:

$\int_a^b y \, dx \approx \frac{h}{2}[(y_0 + y_n) + 2(y_1 + y_2 + ... y_{n-1})]$

where n is the number of strips (in this case 4), and h is the width of each strip:

$h = \frac{b - a}{n} = \frac{2 - 0}{4} = 0.5$ ***[1 mark]***

Work out each y value:

$x_0 = 0 \quad y_0 = 2^{0^2} = 2^0 = 1$

$x_1 = 0.5 \quad y_1 = 2^{0.5^2} = 2^{0.25} = 1.189$ (3 d.p.)

$x_2 = 1 \quad y_2 = 2^{1^2} = 2^1 = 2$

$x_3 = 1.5 \quad y_3 = 2^{1.5^2} = 2^{2.25} = 4.757$ (3 d.p.)

$x_4 = 2 \quad y_4 = 2^{2^2} = 2^4 = 16$ ***[1 mark for all values correct]***

Now put all these values into the formula:

$\int_0^2 2^{x^2} dx \approx \frac{0.5}{2}[(1 + 16) + 2(1.189 + 2 + 4.757)]$ ***[1 mark]***

$= \frac{1}{4}(17 + 15.892) = 8.22$ (3 s.f.) ***[1 mark]***

b) E.g. The curve is concave upwards, so a trapezium on each strip has a greater area than that under the curve. So the trapezium rule gives an overestimate for the area.

[1 mark for a correct answer with an explanation relating to the shape of the graph].

c) To improve the accuracy of the estimate, increase the number of strips used in the trapezium rule ***[1 mark]***.

5 a) When $x = \frac{\pi}{2}$, $y = \frac{\pi}{2} \sin \frac{\pi}{2} = \frac{\pi}{2} = 1.5708$ (5 s.f.),

and when $x = \frac{3\pi}{4}$, $y = \frac{3\pi}{4} \sin \frac{3\pi}{4} = 1.6661$ (5 s.f.)

[1 mark for both]

b) The width of each strip (h) is $\frac{\pi}{4}$, so the trapezium rule gives:

Area of $R \approx \frac{1}{2} \times \frac{\pi}{4}[(0 + 0) + 2(0.55536 + 1.5708 + 1.6661)]$

$= \frac{\pi}{8}[2(3.79226)] = 2.97843... = 2.978$ (4 s.f.)

[3 marks available — 1 mark for correct value of h, 1 mark for using the formula correctly, 1 mark for correct answer]

c) Let $u = x$, so $\frac{du}{dx} = 1$. Let $\frac{dv}{dx} = \sin x$, so $v = -\cos x$
[1 mark for correct choice of u and dv/dx, 1 mark for correct du/dx and v]
Using integration by parts,
$$\int_0^\pi x\sin x\,dx = [-x\cos x]_0^\pi - \int_0^\pi -\cos x\,dx \quad \textbf{\textit{[1 mark]}}$$
$$= [-x\cos x]_0^\pi + [\sin x]_0^\pi \quad \textbf{\textit{[1 mark]}}$$
$$= (-\pi\cos\pi + 0\cos 0) + (\sin\pi - \sin 0) \quad \textbf{\textit{[1 mark]}}$$
$$= (\pi - 0) + (0) = \pi \quad \textbf{\textit{[1 mark]}}$$
If you'd tried to use u = sin x, you'd have ended up with a more complicated function to integrate ($x^2\cos x$).

d) To find the percentage error, divide the difference between the approximate answer and the exact answer by the exact answer and multiply by 100:
$$\left|\frac{\pi - 2.97843...}{\pi}\right| \times 100 = 5.2\%\ (2\text{ s.f.})$$
[2 marks available — 1 mark for appropriate method, 1 mark for correct answer]

6 $n = 5,\ h = \frac{4 - 1.5}{5} = 0.5$
Work out the x- and y-values (y-values given to 5 s.f. where appropriate):
$x_0 = 1.5$ $y_0 = 2.8182$
$x_1 = 2.0$ $y_1 = 4$
$x_2 = 2.5$ $y_2 = 5.1216$
$x_3 = 3.0$ $y_3 = 6.1716$
$x_4 = 3.5$ $y_4 = 7.1364$
$x_5 = 4.0$ $y_5 = 8$
$$\int_{1.5}^{4} y\,dx \approx \frac{0.5}{2}[(2.8182 + 8) + 2(4 + 5.1216 + 6.1716 + 7.1364)]$$
$$= 13.91935 = 13.92 \text{ to 4 s.f.}$$
[4 marks available — 1 mark for the correct value of h, 1 mark for correct x- and y-values, 1 mark for using the formula correctly, 1 mark for the correct answer]

7 a) $n = 5,\ h = \frac{3 - 2.5}{5} = 0.1$
Work out the x- and y-values (y-values given to 6 s.f.):
$x_0 = 2.5$ $y_0 = 0.439820$
$x_1 = 2.6$ $y_1 = 0.424044$
$x_2 = 2.7$ $y_2 = 0.408746$
$x_3 = 2.8$ $y_3 = 0.393987$
$x_4 = 2.9$ $y_4 = 0.379802$
$x_5 = 3.0$ $y_5 = 0.366204$
$$\int_{2.5}^{3} y\,dx \approx \frac{0.1}{2}[(0.439820 + 0.366204) + 2(0.424044 + 0.408746 + 0.393987 + 0.379802)]$$
$$= 0.2009591... = 0.2010\ (4\text{ s.f.})$$
[4 marks available — 1 mark for the correct value of h, 1 mark for correct x- and y-values, 1 mark for using the formula correctly, 1 mark for the correct answer]

b) The area of a rectangle with base 0.5 and height f(2.5) will be an overestimate of area R as the top of the rectangle will be above the top of the curve.
This rectangle has area $0.5 \times 0.439820 = 0.21991$ ***[1 mark]***.
The area of a rectangle with base 0.5 and height f(3) will be an underestimate as the top of the rectangle will be below the top of the curve.
This rectangle has area $0.5 \times 0.366204 = 0.183102$ ***[1 mark]***.
So the actual area of R lies between these two values, i.e. $0.183102 < R < 0.21991$. Both the upper and lower limits round to 0.2, so $R = 0.2$ correct to 1 d.p. ***[1 mark]***

Pages 65-67: Vectors

1 a)

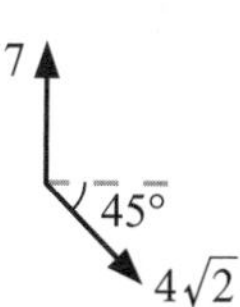

Use trigonometry to find the horizontal and vertical components of the magnitude $4\sqrt{2}$ vector:
Horizontal component: $4\sqrt{2}\cos 315° = 4\mathbf{i}$
Vertical component: $4\sqrt{2}\sin 315° = -4\mathbf{j}$
Resultant, $\mathbf{r} = 7\mathbf{j} + (4\mathbf{i} - 4\mathbf{j})$
So $\mathbf{r} = 4\mathbf{i} + 3\mathbf{j}$
[2 marks available — 1 mark for finding the horizontal and vertical components and 1 mark for correct expression for r]
Remember — angles are normally measured anticlockwise from the positive x-axis, so here, 360° – 45° = 315°

b) $|\mathbf{r}| = \sqrt{4^2 + 3^2} = 5$
$\Rightarrow \mathbf{s} = 7\mathbf{r}$
So $\mathbf{s} = 7(4\mathbf{i} + 3\mathbf{j}) = 28\mathbf{i} + 21\mathbf{j}$
[3 marks available — 1 mark for finding magnitude of r, 1 mark for correct working to find s, 1 mark for correct expression for s]

2 $\overrightarrow{AB} = \overrightarrow{OB} - \overrightarrow{OA} = (5\mathbf{i} - 3\mathbf{j} + 6\mathbf{k}) - (-\mathbf{i} + 7\mathbf{j} - 2\mathbf{k})$
$= 6\mathbf{i} - 10\mathbf{j} + 8\mathbf{k}$ ***[1 mark]***
$\overrightarrow{AM} = \frac{1}{2}\overrightarrow{AB} = 3\mathbf{i} - 5\mathbf{j} + 4\mathbf{k}$
Use this to find $\overrightarrow{CM}$:
$\overrightarrow{CM} = -\overrightarrow{OC} + \overrightarrow{OA} + \overrightarrow{AM}$
$= -(5\mathbf{i} + 4\mathbf{j} + 3\mathbf{k}) + (-\mathbf{i} + 7\mathbf{j} - 2\mathbf{k}) + (3\mathbf{i} - 5\mathbf{j} + 4\mathbf{k})$
$= -3\mathbf{i} - 2\mathbf{j} - \mathbf{k}$
[1 mark for a correct method, 1 mark for the correct vector]
$|\overrightarrow{CM}| = \sqrt{(-3)^2 + (-2)^2 + (-1)^2} = \sqrt{14}$
$|\overrightarrow{AB}| = \sqrt{6^2 + (-10)^2 + 8^2} = \sqrt{200}$
$k = \frac{|\overrightarrow{CM}|}{|\overrightarrow{AB}|} = \frac{\sqrt{14}}{\sqrt{200}}$ ***[1 mark]***
$= \sqrt{\frac{7}{100}} = \frac{1}{10}\sqrt{7}$ ***[1 mark]***

3 $\overrightarrow{AB} = \overrightarrow{OB} - \overrightarrow{OA}$
$= (14\mathbf{i} + 12\mathbf{j} - 9\mathbf{k}) - (-2\mathbf{i} + 4\mathbf{j} - 5\mathbf{k}) = (16\mathbf{i} + 8\mathbf{j} - 4\mathbf{k})$ ***[1 mark]***
$\overrightarrow{AC} = \overrightarrow{OC} - \overrightarrow{OA}$
$= (2\mathbf{i} + \mu\mathbf{j} + \lambda\mathbf{k}) - (-2\mathbf{i} + 4\mathbf{j} - 5\mathbf{k}) = 4\mathbf{i} + (\mu - 4)\mathbf{j} + (\lambda + 5)\mathbf{k}$
[1 mark]
$\overrightarrow{AB}$ and $\overrightarrow{AC}$ both share the point A, so to show they're collinear you need to show that the vectors are parallel i.e. $\overrightarrow{AB} = k\overrightarrow{AC}$ for some constant k. So $(16\mathbf{i} + 8\mathbf{j} - 4\mathbf{k}) = k(4\mathbf{i} + (\mu - 4)\mathbf{j} + (\lambda + 5)\mathbf{k})$
[1 mark]
Equate coefficients of **i**, **j**, **k** separately:
$16 = 4k \Rightarrow k = 4$ ***[1 mark]***
$8 = k(\mu - 4) \Rightarrow \mu = 6$
$-4 = k(\lambda + 5) \Rightarrow \lambda = -6$ ***[1 mark for both μ and λ correct]***

4 $\overrightarrow{AB} = \overrightarrow{OB} - \overrightarrow{OA} = \begin{pmatrix} 4 \\ -12 \\ 8 \end{pmatrix} - \begin{pmatrix} 1 \\ -3 \\ 2 \end{pmatrix} = \begin{pmatrix} 3 \\ -9 \\ 6 \end{pmatrix}$ ***[1 mark]***

D is $\frac{2}{3}$ of the way along $\overrightarrow{AB}$, so $\overrightarrow{AD} = \frac{2}{3}\overrightarrow{AB}$

$= \frac{2}{3}\begin{pmatrix} 3 \\ -9 \\ 6 \end{pmatrix} = \begin{pmatrix} 2 \\ -6 \\ 4 \end{pmatrix}$ ***[1 mark]***

$\overrightarrow{OD} = \overrightarrow{OA} + \overrightarrow{AD} = \begin{pmatrix} 1 \\ -3 \\ 2 \end{pmatrix} + \begin{pmatrix} 2 \\ -6 \\ 4 \end{pmatrix} = \begin{pmatrix} 3 \\ -9 \\ 6 \end{pmatrix}$ ***[1 mark]***

Now, $\overrightarrow{OD} = -\frac{1}{2}\overrightarrow{CE}$

$\Rightarrow \overrightarrow{CE} = -2\overrightarrow{OD} = -2\begin{pmatrix} 3 \\ -9 \\ 6 \end{pmatrix} = \begin{pmatrix} -6 \\ 18 \\ -12 \end{pmatrix}$ ***[1 mark]***

So $\overrightarrow{OE} = \overrightarrow{OC} + \overrightarrow{CE} = \begin{pmatrix} -3 \\ 9 \\ -6 \end{pmatrix} + \begin{pmatrix} -6 \\ 18 \\ -12 \end{pmatrix} = \begin{pmatrix} -9 \\ 27 \\ -18 \end{pmatrix}$ ***[1 mark]***

5 $\overrightarrow{PR} = \overrightarrow{PQ} + \overrightarrow{QR} = \begin{pmatrix} 2 \\ -9 \\ 3 \end{pmatrix} + \begin{pmatrix} 14 \\ 6 \\ 7 \end{pmatrix} = \begin{pmatrix} 16 \\ -3 \\ 10 \end{pmatrix}$ ***[1 mark]***

$|\overrightarrow{PQ}| = \sqrt{2^2 + (-9)^2 + 3^2} = \sqrt{94}$

$|\overrightarrow{QR}| = \sqrt{14^2 + 6^2 + 7^2} = \sqrt{281}$

$|\overrightarrow{PR}| = \sqrt{16^2 + (-3)^2 + 10^2} = \sqrt{365}$

[1 mark for attempting to find magnitudes using Pythagoras, 1 mark for all 3 magnitudes correct]

Find $\angle QPR$ using the cosine rule:

$\cos \angle QPR = \dfrac{94 + 365 - 281}{2 \times \sqrt{94} \times \sqrt{365}} = 0.48048...$ ***[1 mark]***

$\angle QPR = \cos^{-1} 0.48048...$

$= 61.282...° = 61.3°$ (1 d.p.) ***[1 mark]***

6 a) Position vector of drone A:

$\mathbf{a} + \mathbf{b} + \mathbf{c} = (4\mathbf{i} + 6\mathbf{j} + 5\mathbf{k}) + (-\mathbf{i} - 2\mathbf{j} - 2\mathbf{k}) + (-3\mathbf{j} + \mathbf{k})$

$= (3\mathbf{i} + \mathbf{j} + 4\mathbf{k})$ ***[1 mark]***

Position vector of drone B:

$2\mathbf{a} + \mathbf{b} - 3\mathbf{c} = 2(4\mathbf{i} + 6\mathbf{j} + 5\mathbf{k}) + (-\mathbf{i} - 2\mathbf{j} - 2\mathbf{k}) - 3(-3\mathbf{j} + \mathbf{k})$

$= (7\mathbf{i} + 19\mathbf{j} + 5\mathbf{k})$ ***[1 mark]***

So the distance between drones A and B is:

$\sqrt{(7-3)^2 + (19-1)^2 + (5-4)^2}$ ***[1 mark]***

$= \sqrt{341} = 18.466... = 18.5$ m (3 s.f.) ***[1 mark]***

b) Vector from drone A to drone B is:

$(7\mathbf{i} + 19\mathbf{j} + 5\mathbf{k}) - (3\mathbf{i} + \mathbf{j} + 4\mathbf{k}) = 4\mathbf{i} + 18\mathbf{j} + \mathbf{k}$

So the vector to take drone A to 2 m below drone B is:

$(4\mathbf{i} + 18\mathbf{j} + \mathbf{k}) - 2\mathbf{k} = 4\mathbf{i} + 18\mathbf{j} - \mathbf{k}$

[2 marks available — 1 mark for a correct method, 1 mark for the correct answer]

c) The position vector of drone B as it moves in the positive $\mathbf{j}$ direction can be described by:

$(7\mathbf{i} + (19 + \lambda)\mathbf{j} + 5\mathbf{k})$ m for $\lambda \geq 0$ ***[1 mark]***

So at the limit of the drone's range:

$|\, 7\mathbf{i} + (19 + \lambda)\mathbf{j} + 5\mathbf{k}\,| = 50$ m

$\Rightarrow \sqrt{7^2 + (19 + \lambda)^2 + 5^2} = 50$ ***[1 mark]***

$\Rightarrow (19 + \lambda)^2 = 2500 - 49 - 25 = 2426$

$\Rightarrow \lambda^2 + 38\lambda + 361 = 2426$

$\Rightarrow \lambda^2 + 38\lambda - 2065 = 0$ ***[1 mark]***

Solve for λ using the quadratic formula:

$\lambda = \dfrac{-38 \pm \sqrt{38^2 - 4(1)(-2065)}}{2(1)}$ ***[1 mark]***

$\Rightarrow \lambda = \dfrac{-38 \pm \sqrt{9704}}{2}$

$\Rightarrow \lambda = 30.25444...$ (ignore the $-$ve solution as $\lambda \geq 0$)

$\Rightarrow \lambda = 30.3$ (3 s.f) ***[1 mark]***

OK I won't lie, that last part was pretty nasty. But actually... once you've got the initial equation written down, it's just standard algebra that you've been doing since GCSE.

Section Two — Statistics

Pages 68-73: Data Presentation and Interpretation

1 a) sample mean $= \dfrac{\sum x_i}{n} = \dfrac{500}{10} = 50$ ***[1 mark]***

sample variance

$= \dfrac{1}{n-1}\left[\sum x_i^2 - \dfrac{(\sum x_i)^2}{n}\right] = \dfrac{1}{9}\left[25\,622 - \dfrac{500^2}{10}\right]$

$= 69.11...$

So sample standard deviation

$= \sqrt{69.11...} = 8.313... = 8.31$ (3 s.f.) ***[1 mark]***

b) (i) The sample mean will be unchanged ***[1 mark]***, because the new value is equal to the original mean ***[1 mark]***.

(ii) The sample standard deviation will decrease ***[1 mark]***. This is because the sample standard deviation is worked out from the deviation of values from the mean — i.e. $s = \sqrt{\dfrac{1}{n-1}S_{xx}}$, where $S_{xx} = \sum (x_i - \bar{x})^2$. So by adding a new value that's equal to the mean, you're not adding to the total deviation from the mean ($= S_{xx}$), but as you have an extra reading, you now have to divide by a larger value of $(n - 1)$ when you work out the sample standard deviation ***[1 mark]***.

2 The area of the $0 \leq A < 100$ bar is $0.5 \times 20 = 10$ cm^2. The frequency of the $0 \leq A < 100$ class is 50, so 1 cm^2 represents a frequency of $50 \div 10 = 5$ ***[1 mark]***. The frequency of the $500 \leq A < 1000$ class is 15, so it will have an area of 3 cm^2. The class width of the $0 \leq A < 100$ class is 100, and the width for this bar is 0.5 cm, so the width of the bar (in cm) is the width of the class divided by 200. The width of the $500 \leq A < 1000$ bar is $500 \div 200 = 2.5$ cm ***[1 mark]***, and its height is the area divided by the width, so height $= 3 \div 2.5 = 1.2$ cm ***[1 mark]***.

3 a) If the runners were in the 21-25 age group in 2000, then they would be in the 31-35 age group in 2010. The graph shows that the 31-35 age group's mean time to finish the race was higher in 2010 than the 21-25 age group's times in 2000, which would mean the runners were slower, so the graph does not suggest these runners' times improved.

[2 marks available — 1 mark for sensible explanation and 1 mark for the correct conclusion]

b) E.g. That the same runners participated in the 2010 marathon and the 2000 marathon ***[1 mark for a sensible comment]***.

4 Naga's formula will only produce a correct answer if all the different marital-status categories contain the same number of people ***[1 mark]***. This is not the case for this data set, so Naga's formula will not produce a correct result, because it isn't weighted to allow for the different numbers in each category ***[1 mark]***.

5 a) E.g. The data is discrete. / The stem and leaf diagram shows the distribution of the data. / There aren't too many values to display. ***[1 mark for a sensible explanation]***

b) n is even and $\frac{n}{2} = 6$, so take the average of the 6th and 7th values. So the median (Q_2) is:

$\frac{1}{2}(5500 + 5800) = £5650$ ***[1 mark]***.

Since $12 \div 4 = 3$, the lower quartile is the average of the 3rd and 4th values. So the lower quartile (Q_1) is:

$\frac{1}{2}(4200 + 4600) = £4400$ ***[1 mark]***.

Since $12 \div 4 \times 3 = 9$, the upper quartile is the average of the 9th and 10th values. So the upper quartile (Q_3) is:

$\frac{1}{2}(6000 + 6200) = £6100$ ***[1 mark]***.

The interquartile range is $6100 - 4400 = £1700$ ***[1 mark]***.

c) The upper fence for outliers can be defined as $Q_3 + 1.5 \times \text{IQR}$. For this data, $Q_3 + 1.5 \times \text{IQR} = 6100 + 1.5 \times 1700 = £8650$. £9100 is above the upper fence, so it is an outlier.

[1 mark for a sensible answer and calculation]

It's not worth using mean + 2 standard deviations as the fence here as you've already worked out the quartiles and IQR in b).

d) E.g. the manager should not include this outlier in his analysis as it might be caused by an outside factor, e.g. Christmas, so including it would not accurately reflect normal sales. / The manager should include this outlier in his analysis as it accurately reflects actual sales for this period, even if one week was unusually high. ***[1 mark for a sensible comment]***

e) E.g. the mean and standard deviation are affected by outliers, whereas the median and IQR are not. So as this set of data includes an outlier, the median and IQR are more useful. / The mean and standard deviation take into account all data values, so are more useful as they reflect the actual data. ***[2 marks available — 1 mark for stating which values are more useful, 1 mark for a sensible explanation]***

6 a) The $0 \leq b < 10$ class has a width of 10 and a frequency of 12, so its frequency density is $12 \div 10 = 1.2$. Use this to label the vertical axis and draw the missing bars:
$10 \leq b < 15$ class: frequency density $= 23 \div 5 = 4.6$
$25 \leq b < 35$ class: frequency density $= 6 \div 10 = 0.6$

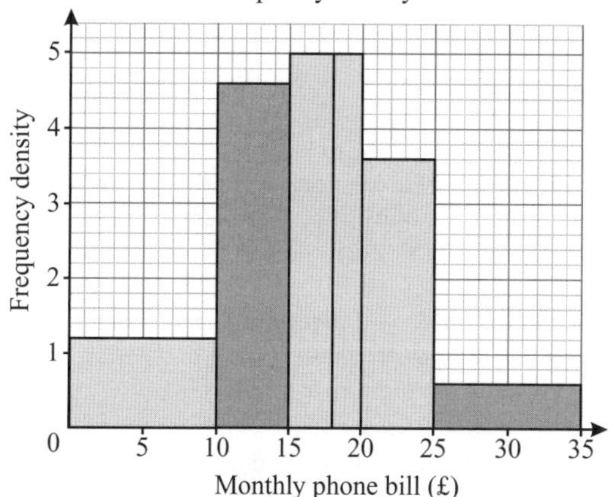

Then use the frequency density axis and the formula frequency = frequency density × class width to work out the missing values in the table:

Monthly phone bill, £*b*	Frequency
$0 \leq b < 10$	12
$10 \leq b < 15$	23
$15 \leq b < 18$	$5 \times 3 = 15$
$18 \leq b < 20$	$5 \times 2 = 10$
$20 \leq b < 25$	18
$25 \leq b < 35$	6

[4 marks available — 1 mark for using the frequency density and the 0 ≤ b < 10 bar or the 20 ≤ b < 25 bar to work out the scale on the vertical axis, 1 mark for both bars correct, 1 mark for using frequency = frequency density × class width, 1 mark for both entries in table correct]

b) £12.50 is halfway through the $10 \leq b < 15$ class, so there are approximately $23 \div 2 = 11.5$ students that pay between £12.50 and £15. £17.50 is $\frac{5}{6}$ of the way through the $15 \leq b < 18$ class, so there are $\frac{5}{6} \times 15 = 12.5$ students that pay between £15 and £17.50. So an estimate for the number of students in the sample who have a monthly phone bill of between £12.50 and £17.50 is $11.5 + 12.5 = 24$. There are 84 students in the sample, so the estimated proportion is $24 \div 84 = 0.2857... = 0.286$ (3 s.f.).
[2 marks available — 1 mark for a correct method, 1 mark for the correct answer]

c) The quickest method for estimating the sample mean and standard deviation is to find the class midpoints, *x*, and to use your calculator functions with the frequencies, *f*, and values of *x*. Or you can work everything out yourself by adding columns to the table and using the formulas:

Class midpoint, x	$f \times x$	x^2	$f \times x^2$
5	60	25	300
12.5	287.5	156.25	3593.75
16.5	247.5	272.25	4083.75
19	190	361	3610
22.5	405	506.25	9112.5
30	180	900	5400
	1370		26 100

There are 84 students in total,
so sample mean $= \frac{\sum fx}{\sum f} = \frac{1370}{84} = 16.309... = £16.31$ (2 d.p.)

Sample standard deviation =

$$\sqrt{\frac{1}{(\sum f)-1} \times \left[\sum fx^2 - \frac{(\sum fx)^2}{\sum f}\right]}$$

$$= \sqrt{\frac{1}{83} \times \left[26\,100 - \frac{1370^2}{84}\right]} = 6.7269... = £6.73 \text{ (2 d.p.)}$$

[3 marks available — 1 mark for correct working, 1 mark for correct mean, 1 mark for correct standard deviation]

d) The 20th percentile is in the $\frac{20}{100} \times 84 = 16.8$th position.
This lies in the class $10 \leq b < 15$.
So 20th percentile $= 10 + 5 \times \frac{16.8 - 12}{23} = 10 + 1.0434...$
$= £11.04$ (2 d.p.)
[3 marks available — 1 mark for calculating the position of the 20th percentile, 1 mark for substituting into the formula for linear interpolation, 1 mark for the correct answer]

7 a) Area A = South West of England
Area B = Barking and Dagenham ***[1 mark for both correct]***
E.g. The line on the graph for Area A shows less variation than the line for Area B, so this will correspond to the area with the smaller standard deviation. / The line for Area A is generally higher than the line for Area B, which corresponds to a higher mean ***[1 mark for a sensible explanation]***.

b) Area A (South West of England):
Outliers are below 186 000 – (2 × 12 808) = £160 384 or above 186 000 + (2 × 12 808) = £211 616 ***[1 mark]***.
Area B (Barking and Dagenham):
Outliers are below 181 800 – (2 × 23 776) = £134 248 or above 181 800 + (2 × 23 776) = £229 352 ***[1 mark]***
So both data sets have an outlier (£215 000 > £211 616 for Area A and £243 500 > £229 352 for Area B) ***[1 mark]***.
If you'd made a mistake in part a) and got the areas the wrong way round, you'd still get the marks for doing the calculations for the outliers.

8 a) The 60th percentile is in the $\frac{60}{100} \times 30 = 18$th position.
So, using the graph, 60th percentile ≈ 4.1 m ***[1 mark]***.

b) E.g. The giraffes in the zoo are generally taller, as they have a higher median. / Two giraffes in the game reserve have extreme heights that are outliers, but there are no outliers in the zoo. / The two populations seem similarly varied, since they have similar ranges (the range for the zoo is 2.9 m and, ignoring the outliers, the range for the game reserve is 2.8 m), although the IQR for the giraffes in the zoo is slightly greater than for the giraffes in the game reserve. / The distribution of heights for the giraffes in the zoo is slightly positively skewed, whereas the distribution of heights for the giraffes in the game reserve is slightly negatively skewed.
[3 marks available — 1 mark for each of three sensible comparisons, one of which should compare the shape of the distributions]

Pages 74-77: Probability

1 a) A and B are mutually exclusive, so $P(A \cap B) = 0$.
This means the circles don't intersect.
B and C are independent, so
$P(B \cap C) = P(B) \times P(C) = 0.4 \times 0.3 = 0.12$
$P(A \cap C) = 0.06$
Use this to fill in the Venn diagram:

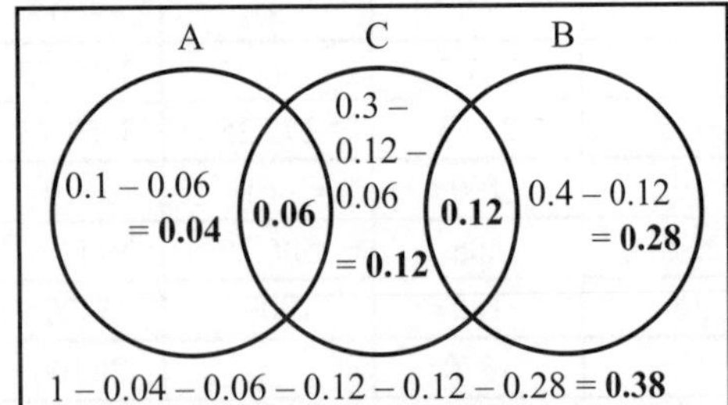

[5 marks available — 1 mark for circles overlapping correctly, 1 mark for calculating P(B ∩ C), 1 mark for calculating the remaining probabilities in both circles A and B, 1 mark for calculating the remaining probability in circle C, 1 mark for the correct probability outside the circles]

b) $P(A) \times P(C) = 0.1 \times 0.3 = 0.03$
$P(A \cap C) = 0.06 \neq P(A) \times P(C)$,
so events A and C are not independent.
[2 marks available — 1 mark for finding P(A) × P(C) and stating P(A ∩ C), 1 mark for the correct conclusion]

c) $P(B \cup C) = 0.06 + 0.12 + 0.12 + 0.28 = 0.58$ ***[1 mark]***

d) $P(A' \cap B') = 0.12 + 0.38 = 0.5$ ***[1 mark]***

e) $P(B'|A') = \frac{P(B' \cap A')}{P(A')} = \frac{0.5}{0.9} = \frac{5}{9} = 0.56$ (2 d.p.)
[2 marks available — 1 mark for using the conditional probability formula, 1 mark for the correct answer]
Use the numbers in the Venn diagram to answer parts c)-e).

2 As L and M are independent, $P(L \cap M) = P(L) \times P(M)$ ***[1 mark]***
$P(L) = 0.1 + 0.4 = 0.5$ and $P(M) = 2x + 0.4$ ***[1 mark]***
$P(L \cap M) = 0.4 = 0.5 \times (2x + 0.4) = x + 0.2 \Rightarrow x = 0.2$ ***[1 mark]***
So this means $P(L \cup M) = 0.1 + 0.4 + 0.4 = 0.9$
$\Rightarrow P(L' \cap M') = 1 - 0.9 = 0.1$ ***[1 mark]***

3 Draw a tree diagram:

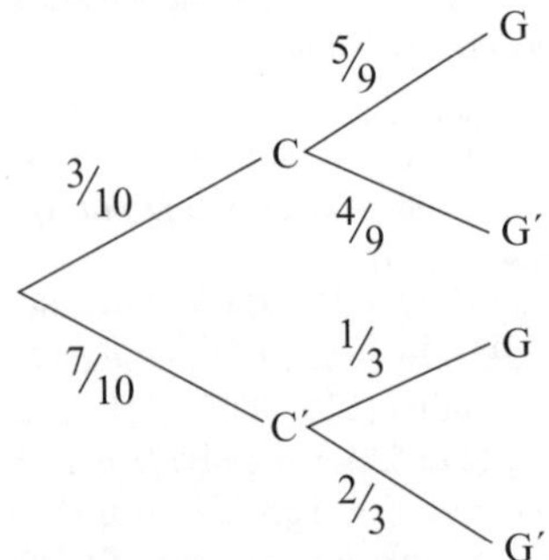

You need to find: $P(C|G) = \frac{P(C \cap G)}{P(G)}$ ***[1 mark]***.
$P(G) = P(C \cap G) + P(C' \cap G)$:
$P(C \cap G) = \frac{3}{10} \times \frac{5}{9} = \frac{15}{90} = \frac{1}{6}$ ***[1 mark]***,
and $P(C' \cap G) = \frac{7}{10} \times \frac{1}{3} = \frac{7}{30}$.
Adding these together you get: $P(G) = \frac{1}{6} + \frac{7}{30} = \frac{2}{5}$ ***[1 mark]***.
So $P(C|G) = \frac{1}{6} \div \frac{2}{5} = \frac{5}{12}$ ***[1 mark]***.

4 a) Let U and M be the events 'selected fan is under 18' and 'selected fan is male'. There are 1945 males, of which 305 are under 18,
so $P(U | M) = \frac{305}{1945} = \frac{61}{389}$.
[2 marks available — 1 mark for a correct method, 1 mark for the correct answer]
You could have found this using the formula for conditional probability.

b) E.g. Alan has assumed that the fans at the next match will have the same mix of ages and genders, but many factors affect the attendance at a game, and different opponents may have a different age or gender balance in their fans.
[2 marks available — 1 mark for an assumption, 1 mark for a sensible reason why this may not be valid]

5 a) Draw a tree diagram. Let C = 'pick a heart from the complete pack':

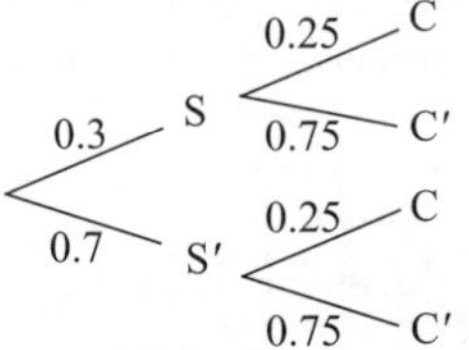

The probability of at least one event happening is the same as 1 minus the probability of neither event happening:
$P(S \cup C) = 1 - P(S' \cap C')$
$= 1 - (0.7 \times 0.75) = 1 - 0.525 = 0.475$
[2 marks available — 1 mark for a correct method, 1 mark for the correct answer]
You could also get the answer by doing $P(S \cap C) + P(S \cap C') + P(S' \cap C)$.

b) Let P(H) be the probability of drawing at least two hearts, and let C_1, C_2 and C_3 be the event of selecting a heart with the first, second and third card respectively.
Draw a tree diagram to calculate the probability of $P(H \cap C_1)$:

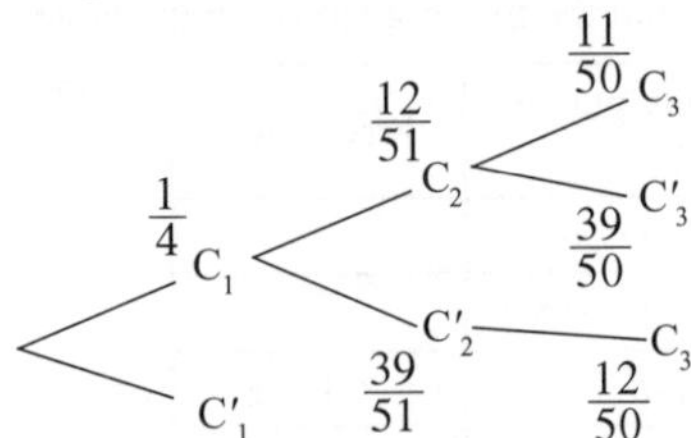

The branches of this tree diagram that are labelled with probabilities are the only ones where both a heart is selected first and at least two hearts are selected, so:
$$P(H \cap C_1) = \left(\frac{1}{4} \times \frac{12}{51} \times \frac{11}{50}\right) + \left(\frac{1}{4} \times \frac{12}{51} \times \frac{39}{50}\right) + \left(\frac{1}{4} \times \frac{39}{51} \times \frac{12}{50}\right) = \frac{89}{850}$$
So, $P(H | C_1) = \frac{P(H \cap C_1)}{P(C_1)} = \frac{89}{850} \div \frac{1}{4} = \frac{356}{850}$
$= 0.419$ (3 s.f.)
[3 marks available — 1 mark for a correct method to find P(H ∩ C₁), 1 mark for P(H ∩ C₁) correct and 1 mark for the correct answer]
You could also get this answer without using the formula, by just considering the tree diagram from the point after C_1 has happened.

c) E.g. I have assumed that Jessica is equally likely to select any of the cards in the complete pack of cards.
[1 mark for any sensible comment]

6 $P(A | B) = 0.31 \Rightarrow \frac{P(A \cap B)}{0.4} = 0.31 \Rightarrow P(A \cap B) = 0.124$
[1 mark]
Now substitute this into the formula for $P(B | A)$:
$P(B | A) = \frac{0.124}{P(A)} = 0.25 \Rightarrow P(A) = \frac{0.124}{0.25} = 0.496$ ***[1 mark]***
So, $P(A') = 1 - 0.496 = 0.504$ ***[1 mark]***

7 a) Draw a Venn diagram using the information you know:

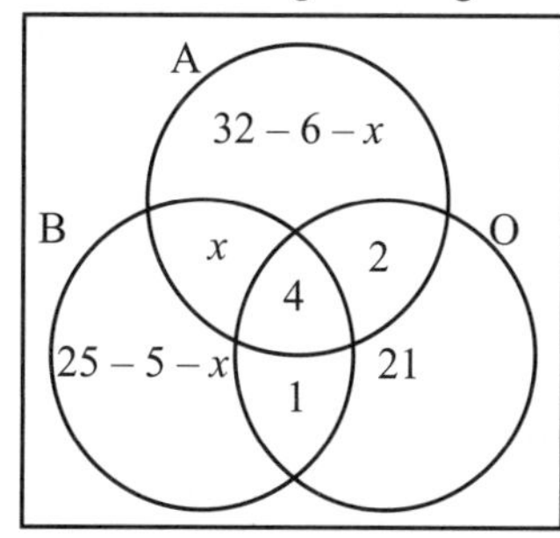

[1 mark]

Everything inside the circles should add up to 65:
$(32 - 6 - x) + x + 4 + 2 + (25 - 5 - x) + 1 + 21 = 65$ ***[1 mark]***
$74 - x = 65 \Rightarrow x = 9$ ***[1 mark]***
So $x + 4 = 13$ children took an apple and a banana ***[1 mark]***

b) Draw a tree diagram (X is the first child selects an apple, Y is the second child selects an apple):

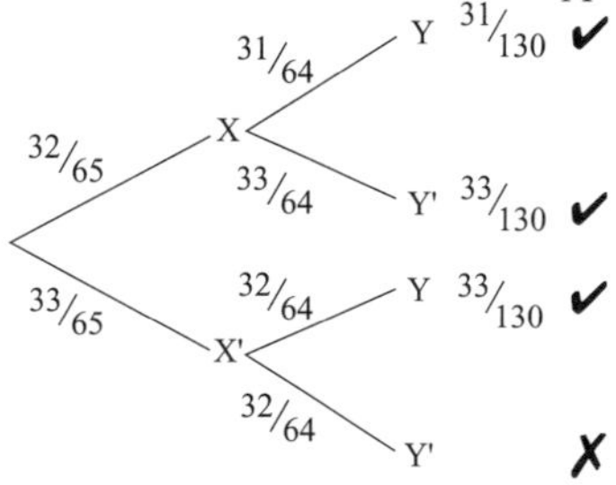

The ticked outcomes are those where at least one person takes an apple. So $P(X \cup Y) = \frac{31}{130} + \frac{33}{130} + \frac{33}{130} = \frac{97}{130}$ ***[1 mark]***

$P(X | X \cup Y) = \frac{P(X \cap (X \cup Y))}{P(X \cup Y)}$ ***[1 mark]***

As all of X is contained within $X \cup Y$, $X \cap (X \cup Y)$ will be all of X. ***[1 mark]***

So $\frac{P(X \cap (X \cup Y))}{P(X \cup Y)} = \frac{P(X)}{P(X \cup Y)} = \frac{32}{65} \div \frac{97}{130} = \frac{64}{97}$ ***[1 mark]***

Pages 78-84: Statistical Distributions

1 a) The total probability must be 1, so go through all the possible values of x and add the probabilities:

$\frac{k}{6} + \frac{2k}{6} + \frac{3k}{6} + \frac{3k}{6} + \frac{2k}{6} + \frac{k}{6} = \frac{12k}{6} = 2k$ ***[1 mark]***.

This must equal 1, so k must be $\frac{1}{2}$ ***[1 mark]***.

b) $P(1 < X \leq 4) = P(X = 2) + P(X = 3) + P(X = 4)$ ***[1 mark]***

$= \frac{1}{6} + \frac{1}{4} + \frac{1}{4} = \frac{2}{3}$ ***[1 mark]***

c) Every value of y has the same probability, and there are no other possible values as these add up to 1, so Y follows a discrete uniform distribution ***[1 mark]***.

2 Using $\sum_{\text{all } x} P(X = x) = 1$:

$0.4 + 0.3 + a + b = 1$ ***[1 mark]*** $\Rightarrow 0.7 + a + b = 1 \Rightarrow a + b = 0.3$

A contestant is twice as likely to be awarded 2 points as 3 points, so you also know that $a = 2b$. Substitute this in the above to get:

$2b + b = 0.3 \Rightarrow 3b = 0.3 \Rightarrow b = 0.1$ and $a = 2b = 0.2$

[1 mark for both correct]

P(one contestant scores 2 and the other scores 3)
= P(1st scores 2 and 2nd scores 3 or 1st scores 3 and 2nd scores 2)
$= (0.2 \times 0.1) + (0.1 \times 0.2)$ ***[1 mark]***
$= 0.04$ ***[1 mark]***

3 a) The probability of getting 3 heads is:

$\frac{1}{2} \times \frac{1}{2} \times \frac{1}{2} = \frac{1}{8}$

The probability of getting 2 heads is: $3 \times \frac{1}{2} \times \frac{1}{2} \times \frac{1}{2} = \frac{3}{8}$

(multiply by 3 because any of the three coins could be the tail — the order in which the heads and the tail occur isn't important).

The probability of getting 1 or 0 heads
$= 1 - P(2 \text{ heads}) - P(3 \text{ heads}) = 1 - \frac{3}{8} - \frac{1}{8} = \frac{1}{2}$

Hence the probability distribution of X is:

x	20p	10p	nothing
$P(X = x)$	$\frac{1}{8}$	$\frac{3}{8}$	$\frac{1}{2}$

[3 marks for a completely correct table, otherwise 1 mark for drawing a table with the correct values of x and 1 mark for any one correct probability]

b) There are three ways to make a profit over two games: win 20p and 10p, win 10p and 20p, or win 20p and 20p.

So P(makes profit over 2 games)

$= \left(\frac{1}{8} \times \frac{3}{8}\right) + \left(\frac{3}{8} \times \frac{1}{8}\right) + \left(\frac{1}{8} \times \frac{1}{8}\right)$

$= \frac{3}{64} + \frac{3}{64} + \frac{1}{64} = \frac{7}{64}$ ***[1 mark]***

$P(\text{wins 40p}) = \frac{1}{8} \times \frac{1}{8} = \frac{1}{64}$ ***[1 mark]***

P(wins 40p | makes profit over two games)

$= \frac{P(\text{wins 40p and makes profit})}{P(\text{makes profit})} = \frac{P(\text{wins 40p})}{P(\text{makes profit})}$ ***[1 mark]***

$= \frac{1/64}{7/64} = \frac{1}{7}$ ***[1 mark]***

4 a) 1) The probability P(chocolate bar contains a golden ticket) must be constant.

2) The trials must be independent (i.e. whether or not a chocolate bar contains a golden ticket must be independent of other chocolate bars). ***[1 mark for each correct condition]***

b) $P(X > 1) = 1 - P(X \leq 1)$ ***[1 mark]***

$= 1 - 0.3990... = 0.6009$ (4 d.p.) ***[1 mark]***.

c) Mean of $X = np = 40 \times 0.05 = 2$, so the student can expect to find 2 golden tickets.

[1 mark for the correct answer with justification]

5 a) Let the random variable X represent the number of cars in the sample of 20 that develop the rattle.

Then $X \sim B(20, 0.65)$ ***[1 mark]***,
and you need to find $P(12 \leq X < 15)$.
$P(12 \leq X < 15) = P(X \leq 14) - P(X \leq 11)$
$= 0.7546... - 0.2376...$ ***[1 mark for either]***
$= 0.5170$ (4 d.p.) ***[1 mark]***

b) $P(X > 10) = 1 - P(X \leq 10)$ ***[1 mark]***
$= 1 - 0.1217...$
$= 0.8782$ (4 d.p.) ***[1 mark]***.

c) The probability of more than half of a sample of 20 cars having the rattle is 0.8782 (from part b).

Let Q be the number of samples containing more than 10 rattling cars. Then $Q \sim B(5, 0.8782)$ ***[1 mark]***.

$P(Q = 3) = \binom{5}{3} \times 0.8782^3 \times (1 - 0.8782)^2$ ***[1 mark]***

$= 0.1005$ (to 4 d.p.) ***[1 mark]***

You could use your calculator's binomial distribution functions to find P(Q = 3).

6 a) Let X be the number of customers in the sample of 20 who chose a sugar cone. Then $X \sim B(n, p)$ where $n = 20$, but you need to find p (the probability of choosing a sugar cone).

p is equal to the proportion of customers on the 1st July who chose a sugar cone, so:

$p = \frac{880}{1100} = 0.8$ ***[1 mark]***, so $X \sim B(20, 0.8)$ ***[1 mark]***

$P(X = 12) = \binom{20}{12} \times 0.8^{12} \times 0.2^8$

$= 0.0222$ (4 d.p.) ***[1 mark]***

Or you could use your calculator's binomial distribution functions here.

b) Let Y be the number of customers who choose at least one scoop of chocolate ice cream.

Then $Y \sim B(75, 0.42)$ ***[1 mark]***.
$P(Y > 30) = 1 - P(Y \leq 30)$ ***[1 mark]***
$= 1 - 0.4099... = 0.5901$ (4 d.p.) ***[1 mark]***

You need to use your calculator c.d.f. for this one.

c) E.g. it is not reasonable to assume that the customers' choices of ice cream are independent of each other, e.g. one person might see another person with a chocolate ice cream and decide to get one too, so the 75 trials are not independent. / The probability of choosing chocolate ice cream may not be constant, e.g. it might run out. Therefore the model may not be valid ***[1 mark for a sensible comment]***.

7 a) E.g. The model is appropriate because (any two of):
- The histogram is roughly the shape of a normal distribution (i.e. bell-shaped).
- The histogram is roughly symmetrical about the mean.
- All the data lies between 2.0 kg and 5.0 kg (and 3 standard deviations from the mean is the range 2.05-5.05 kg).

[2 marks for stating the model is appropriate and giving two sensible reasons, otherwise 1 mark for a sensible comment]

b) Let M represent the masses of the newborn babies.
$M \sim N(3.55, 0.5^2)$
$P(M < 2.5) = 0.0179$ (4 d.p.)
[2 marks available — 1 mark for setting up the calculation correctly, 1 mark for the correct answer]

8 a) The mean lies halfway between the points of inflection,
so $\mu = \frac{5940 + 6030}{2} = 5985$ ml ***[1 mark]***
The points of inflection occur at $\mu \pm \sigma$,
so $6030 = 5985 + \sigma \Rightarrow \sigma = 45$ ml ***[1 mark]***
You could have found μ and σ by setting up and solving two simultaneous equations ($\mu + \sigma = 6030$ and $\mu - \sigma = 5940$), or by finding σ first.

b) $v \sim N(5985, 45^2)$
$P(5900 \le v \le 6100) = 0.9652$ (4 d.p.)
[2 marks available — 1 mark for setting up the calculation correctly, 1 mark for the correct answer]

9 a) Let X represent the exam marks. Then $X \sim N(50, 15^2)$.
$P(X < 30) = 0.0912$ (4 d.p.)
[2 marks available — 1 mark for setting up the calculation correctly, 1 mark for the correct answer]

b) $P(X \ge 41) = 0.7257$ (4 d.p.) ***[1 mark]***
$0.7257 \times 1000 = 725.7 \approx 726$ candidates passed the exam ***[1 mark]***

c) You need to find x such that $P(X \ge x) = 0.1$
Using the inverse normal function,
$x = 69$ marks (to the nearest whole number).
[2 marks available — 1 mark for a correct calculation, 1 mark for the correct answer]
You need to use your calculator's inverse normal function here.

10 a) Let X represent the base diameters.
Then $X \sim N(12, \sigma^2)$. $P(X > 13) = 0.05$ ***[1 mark]***,
so $P\left(Z > \frac{13 - 12}{\sigma}\right) = P\left(Z > \frac{1}{\sigma}\right) = 0.05$
Using the percentage points table, $p = 10$, so $\frac{1}{2}p = 5$ for $z = 1.645$
$\Rightarrow \frac{1}{\sigma} = 1.645$ ***[1 mark]***
$\Rightarrow \sigma = 1 \div 1.645 = 0.608$ inches (to 3 s.f.) ***[1 mark]***

b) $X \sim (12, 0.608^2)$
$P(X < 10.8) = 0.0242$ (4 d.p.) ***[1 mark for setting up the calculation, 1 mark for the answer]***
So you would expect $0.0242 \times 100 \approx 2$ pizza bases to be discarded ***[1 mark]***.

c) P(at least 1 base too small) = 1 – P(no bases too small)
P(base not too small) = 1 – 0.0242 = 0.9758
P(no bases too small) = $0.9758^3 = 0.9291$ ***[1 mark]***
P(at least 1 base too small) = 1 – 0.9291 ***[1 mark]***
= 0.0709 (4 d.p.) ***[1 mark]***

11 a) $X \sim N(4, 1.1^2)$
$P(X < 3.5) = 0.3247$ (4 d.p.)
[2 marks available — 1 mark for setting up the calculation correctly, 1 mark for the correct answer]

b) 'Deviates from the mean by more than 1 minute' means the window cleaner takes less than 4 – 1 = 3 minutes or more than 4 + 1 = 5 minutes, so you need to find:
$P(X < 3) + P(X > 5) = 0.1816... + 0.1816... = 0.3633$ (4 d.p.)
[3 marks available — 1 mark for finding the values either side of the mean, 1 mark for setting up the calculation correctly, 1 mark for the correct answer]
Here, you could have worked out $1 - P(3 < X < 5)$ instead.

c) You need to find t such that $P(X > t) = 0.01$
Using the inverse normal function, $t = 6.6$ mins (1 d.p.) (in minutes and seconds, this is 6 min and 36 seconds).
[2 marks available — 1 mark for a correct calculation, 1 mark for the correct answer]
If your calculator doesn't have an inverse normal function, you can convert to the standard normal distribution here and use the percentage points table — look up $p = 2$ to give you z such that $P(Z > z) = 0.01$.

12 $M \sim N(8.2, 0.6)$ and $K \approx 1.6M$,
so $K \sim N(1.6 \times 8.2, 1.6^2 \times 0.6) = N(13.12, 1.536)$.
[2 marks for a normal distribution with the correct mean and variance, otherwise 1 mark for either the mean or variance correct]

13 a) X represents the number of sampled customers at Soutergate Cinema who went to see the superhero film,
so $X \sim B(200, 0.48)$ ***[1 mark]***.

b) A normal approximation is suitable here because the shape of the distribution of X is symmetrical, so a normal distribution would be a close approximation.
[1 mark for a sensible comment based on the graph]

c) The distribution of X is discrete, but the normal distribution is for continuous data, so a continuity correction is needed ***[1 mark]***.

d) $X \sim B(200, 0.48)$, so the mean of $X = 200 \times 0.48 = 96$ ***[1 mark]***
Variance = $npq = 200 \times 0.48 \times (1 - 0.48) = 49.92$ ***[1 mark]***
So $X \sim B(200, 0.48)$ can be approximated by $Y \sim N(96, 49.92)$.

Pages 85-88: Statistical Hypothesis Testing

1 a) All the pupils in her school ***[1 mark]***.

b) Opportunity sampling ***[1 mark]***
It is unlikely to be representative because, e.g: the members of the sample are all in the same age group, so won't represent the whole school / older pupils are likely to have different views on politics to younger pupils / A-level politics students might be biased towards certain views / all the pupils who aren't in Josie's A-level politics class are excluded from the sample.
[1 mark for a sensible explanation]

c) E.g. Josie will only get responses from those pupils who choose to reply to her email / Those who make the effort to reply may be more likely to have strong views.
[1 mark for a sensible explanation]

2 a) E.g. people in different age groups may be likely to receive higher or lower pay rises, so it is important that the proportion in each age group in the sample is representative of the proportion in the population.
[1 mark for a sensible explanation]

b) Total population of working adults
= 1200 + 2100 + 3500 + 3200 + 1500 = 11 500

18-27 years: $\frac{1200}{11\,500} \times 50 = 5.217... \approx 5$

28-37 years: $\frac{2100}{11\,500} \times 50 = 9.130... \approx 9$

38-47 years: $\frac{3500}{11\,500} \times 50 = 15.217... \approx 15$

48-57 years: $\frac{3200}{11\,500} \times 50 = 13.913... \approx 14$

Over 57 years: $\frac{1500}{11\,500} \times 50 = 6.521... \approx 7$

[3 marks available — 1 mark for using the correct total in the calculations, 1 mark for at least 2 correct values, 1 mark for all 5 correct values]

Section Three — Mechanics

Pages 92-96: Kinematics 1

1 Displacement (S) is given by the area under the graph. The area under the line is the area of a trapezium:

area under the line $= \frac{U+V}{2} \times T$

So $S = \frac{1}{2}(U+V)T$, as required.

[3 marks available — 1 mark for stating that displacement is the area under the graph, 1 mark for a using a suitable method to find the area under the line, 1 mark for the correct conclusion]

2 a) $u = 5, v = 0, a = -9.8, t = ?$

Using $v = u + at$: $0 = 5 - 9.8t \Rightarrow t = 5 \div 9.8 = 0.5102...$
$t = 0.510$ s (3 s.f.)

[3 marks available — 1 mark for using appropriate equation, 1 mark for correct workings, 1 mark for correct value of t]

b) $s = -2, u = 5, v = ?, a = -9.8$ Using $v^2 = u^2 + 2as$:

$v^2 = 5^2 + 2(-9.8 \times -2) = 64.2 \Rightarrow v = -8.0124...$
$= -8.01\ \text{ms}^{-1}$ (3 s.f.)

The velocity is negative when it reaches B as the ball is travelling in the negative direction (you can also see this from the graph).

[3 marks available — 1 mark for using appropriate equation, 1 mark for correct workings, 1 mark for correct value of v]

c) E.g. The graph shows the ball instantaneously changing velocity from point B to point C, which is unrealistic.

[1 mark for any sensible comment]

3 From passing the sign to entering the tunnel:

$s = ut + \frac{1}{2}at^2 \Rightarrow 110 = 8u + \frac{1}{2}(a \times 8^2)$ ***[1 mark]***

$\Rightarrow 110 = 8u + 32a$ ① ***[1 mark]***

From passing the sign to leaving the tunnel:

$s = ut + \frac{1}{2}at^2 \Rightarrow 870 = 32u + \frac{1}{2}(a \times 32^2)$ ***[1 mark]***

$\Rightarrow 870 = 32u + 512a$ ② ***[1 mark]***

② – 4 × ① gives: ***[1 mark for a correct method]***

$430 = 384a \Rightarrow a = 1.119... = 1.12\ \text{ms}^{-2}$ (3 s.f.) ***[1 mark]***

Substitute back into ① to find u:

$110 = 8u + 32a$

$u = \frac{110 - 32(1.119...)}{8} = 9.270... = 9.27\ \text{ms}^{-1}$ ***[1 mark]***

4 a) $|\sqrt{1^2 + 3^2} - \sqrt{(-5)^2 + 2^2}| = |\sqrt{10} - \sqrt{29}| = 2.22\ \text{ms}^{-1}$ (3 s.f.)

[2 marks available — 1 mark for attempting to find the magnitude of each vector and 1 mark for the correct answer]

b) Draw a diagram:

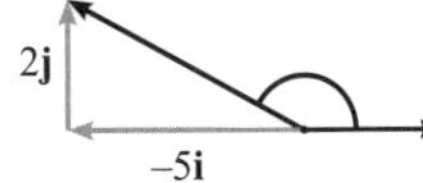

So, angle is $180° - \tan^{-1}\left(\frac{2}{5}\right) = 158.2°$ (1 d.p.)

[2 marks available — 1 mark for attempting to find the direction of the vector and 1 mark for the correct answer]

c) Using $\mathbf{s} = \mathbf{v}t$, the position vector of A is:

$\mathbf{s} = (2\mathbf{i} + \mathbf{j}) + 8(\mathbf{i} + 3\mathbf{j}) = (2\mathbf{i} + \mathbf{j}) + (8\mathbf{i} + 24\mathbf{j})$
$= (10\mathbf{i} + 25\mathbf{j})$ m ***[1 mark]***

Using $\mathbf{s} = \mathbf{v}t$, the position vector of B is:

$\mathbf{s} = (10\mathbf{i} + 25\mathbf{j}) + 5(-5\mathbf{i} + 2\mathbf{j}) = (10\mathbf{i} + 25\mathbf{j}) + (-25\mathbf{i} + 10\mathbf{j})$
$= (-15\mathbf{i} + 35\mathbf{j})$ m ***[1 mark]***

5 a) Find the acceleration using the displacement at $t = 3$:

$\mathbf{u} = \mathbf{0}, \mathbf{a} = \mathbf{a}, \mathbf{s} = (9\mathbf{i} - 18\mathbf{j}), t = 3$

$\mathbf{s} = \mathbf{u}t + \frac{1}{2}\mathbf{a}t^2$

$(9\mathbf{i} - 18\mathbf{j}) = \frac{1}{2} \times \mathbf{a} \times 3^2$ ***[1 mark]***

$(9\mathbf{i} - 18\mathbf{j}) = 4.5\mathbf{a} \Rightarrow \mathbf{a} = (2\mathbf{i} - 4\mathbf{j})\ \text{ms}^{-2}$ ***[1 mark]***

The particle starts at O, so its position vector at time t is equal to its displacement:

So $\mathbf{p} = \mathbf{s} = \frac{1}{2}(2\mathbf{i} - 4\mathbf{j})t^2$ ***[1 mark]***

$\mathbf{p} = (\mathbf{i} - 2\mathbf{j})t^2$ m ***[1 mark]***

b) Find an expression for the second particle's displacement after t seconds:

$\mathbf{s} = \mathbf{v}t$

$\mathbf{s} = (3\mathbf{i} - 5\mathbf{j})t$ ***[1 mark]***

So the second particle's position vector at time t is:

$\mathbf{q} = (a\mathbf{i} + b\mathbf{j}) + (3\mathbf{i} - 5\mathbf{j})t$

When they collide, their position vectors are equal:

$(\mathbf{i} - 2\mathbf{j})t^2 = (a\mathbf{i} + b\mathbf{j}) + (3\mathbf{i} - 5\mathbf{j})t$

$t = 8$, so:

$64(\mathbf{i} - 2\mathbf{j}) = (a\mathbf{i} + b\mathbf{j}) + 8(3\mathbf{i} - 5\mathbf{j})$ ***[1 mark]***

$64\mathbf{i} - 128\mathbf{j} - 24\mathbf{i} + 40\mathbf{j} = a\mathbf{i} + b\mathbf{j}$

$40\mathbf{i} - 88\mathbf{j} = a\mathbf{i} + b\mathbf{j}$

So $a = 40$ and $b = -88$ ***[1 mark for both]***

6 a) The ball is travelling eastwards when the **j** component of the velocity vector is 0 and the **i** component is positive.

$3t - 8 = 0 \Rightarrow t = \frac{8}{3}$.

When $t = \frac{8}{3}$, the **i** component is $2\left(\frac{8}{3}\right) - 5 = \frac{1}{3} > 0$ ***[1 mark]***, so the ball is travelling eastwards at $t = \frac{8}{3}$ seconds ***[1 mark]***.

b) Integrate **v** to find an expression for the ball's horizontal position (ignoring the **j** part of the calculation) :

$\int \mathbf{v}_i\, dt = \mathbf{s}_i = \int (2t - 5)\mathbf{i}\, dt$ ***[1 mark]***
$= (t^2 - 5t)\mathbf{i} + \mathbf{c}$ ***[1 mark]***

When $t = 2$, $\mathbf{s} = -2\mathbf{i} + 10\mathbf{j}$. Use this to find **c**:

$(4 - 10)\mathbf{i} + \mathbf{c} = -2\mathbf{i} \Rightarrow \mathbf{c} = 4\mathbf{i}$ ***[1 mark]***

So, $\mathbf{s}_i = (t^2 - 5t + 4)\mathbf{i}$.

The ball is west of the origin whenever

$t^2 - 5t + 4 = (t - 1)(t - 4) < 0$ ***[1 mark]*** $\Rightarrow 1 < t < 4$. ***[1 mark]***

7 a) $s = \int v\, dt = \frac{11}{2}t^2 - \frac{2}{3}t^3 + c$ for $0 \le t \le 5$.

When $t = 0$, $s = 0 \Rightarrow c = 0$, so $s = \frac{11}{2}t^2 - \frac{2}{3}t^3$

So when $t = 5$:

$s = \frac{11}{2}(25) - \frac{2}{3}(125) = 54.16... = 54.2$ m (3 s.f.)

[4 marks available — 1 mark for integrating with respect to time, 1 mark for obtaining an expression for s, 1 mark for finding c, 1 mark for correct final answer]

b) $s = \int v\, dt$ ***[1 mark]***

$= 25t - 2t^2 + k$ for $t > 5$ ***[1 mark]***

When $t = 5$, $s = 54.16...$ (from part a)) — use this to find k:

$54.16... = 25(5) - 2(25) + k$. So, $k = -20.83...$ ***[1 mark]***

Need to find t when $s = 0$, i.e. when:

$2t^2 - 25t + 20.83... = 0$ ***[1 mark]***

Solve using quadratic formula: $t = 0.898$ or 11.6 s

$t = 0.898$ can be ignored, as it is outside the interval for which the equation is valid.

So P is back at the origin after 11.6 s (3 s.f.) ***[1 mark]***

8 Integrate the acceleration to find an expression for the velocity.

a is a constant, so $v = \int a\, dt = at + c$ ***[1 mark]***.

When $t = 0$, $v = u$, so $u = c$ and this gives $v = at + u$ ***[1 mark]***.

Integrate again to find an expression for the object's displacement:

$s = \int v\, dt = \int at + u\, dt = \frac{1}{2}at^2 + ut + k$ ***[1 mark]***.

The object sets off from the origin, so setting $t = 0$, $s = 0 \Rightarrow k = 0$ which gives $s = ut + \frac{1}{2}at^2$ as required ***[1 mark]***.

9 a) $a = \frac{dv}{dt}$ ***[1 mark]***

$a = 2 - 3(-2)e^{-2t} = (2 + 6e^{-2t})\ \text{ms}^{-2}$ ***[1 mark]***

b) t must be greater than or equal to 0.

When $t = 0$, $a = 2 + 6e^{-2\times0} = 8\ \text{ms}^{-2}$ ***[1 mark]***

As $t \to \infty$, $6e^{-2t} \to 0$, so $a \to 2\ \text{ms}^{-2}$ ***[1 mark]***

So, $2 < a \le 8$ ***[1 mark]***

c) $s = \int v\,dt$ ***[1 mark]***

$s = t^2 + \frac{3}{2}e^{-2t} + 4t + c$ ***[1 mark]***

When $t = 0$, the particle is at the origin (i.e. $s = 0$):

$0 = 0 + \frac{3}{2} + c \Rightarrow c = -\frac{3}{2}$

So, $s = \left(t^2 + \frac{3}{2}e^{-2t} + 4t - \frac{3}{2}\right)$ m ***[1 mark]***

10 a) $\mathbf{a} = \dot{\mathbf{v}} = [(-6\sin 3t + 5)\mathbf{i} + 2\mathbf{j}]$ ms^{-2}

[2 marks in total — 1 mark for attempting to differentiate the velocity vector, 1 mark for correctly differentiating both components]

b) The **j**-component of the acceleration is constant, so the **i**-component is the only one which can be maximised.

To do this, set $\sin 3t = -1$.

*The minimum value of sin(anything) is −1 and because you're multiplying it by −6, this will give you the maximum value for the **i**-component.*

$\mathbf{a} = ((-6 \times -1) + 5)\mathbf{i} + 2\mathbf{j} = 11\mathbf{i} + 2\mathbf{j}$ ***[1 mark]***

Magnitude $= |\mathbf{a}| = \sqrt{11^2 + 2^2}$ ***[1 mark]*** $= \sqrt{121 + 4}$

$= \sqrt{125} = 5\sqrt{5}$ ms^{-2} ***[1 mark]***

Pages 97-98: Kinematics 2

1 a) Consider vertical motion, taking up as positive:

$s = 0;\ u = p;\ a = -9.8;\ t = 5;$

Acceleration is constant, so can use a constant acceleration formula, e.g. use $s = ut + \frac{1}{2}at^2$

$0 = (5p) - 4.9(5^2)$ ***[1 mark for using a suvat equation correctly]***

$5p = 122.5$

$p = 24.5$ ***[1 mark]***

You could also use v = u + at here, with v = 0; u = p; a = −9.8; and t = 2.5 (i.e. the time taken for the ball to reach its highest point, where vertical velocity is momentarily zero).

b) No acceleration horizontally, so use $s = ut$ ***[1 mark]***

$s = 29 \times 5 = 145$ m ***[1 mark]***

2 a) Need to find the times when the stone is 22 m above the ground (i.e. when it is at the level of projection), so consider vertical motion, taking up as +ve: ***[1 mark]***

$s = 0;\ u = 14\sin 46°;\ a = -9.8;\ t = ?$

Acceleration is constant, so use a constant acceleration formula, e.g. $s = ut + \frac{1}{2}at^2$

$0 = 14\sin 46t° - 4.9t^2$ ***[1 mark]***

$0 = (14\sin 46° - 4.9t)t$

So, $t = 0$ (i.e. when the stone is thrown) or

$14\sin 46° - 4.9t = 0 \Rightarrow t = 2.055$ s ***[1 mark]***

So the stone is at least 22 m above the ground for:

$2.055 - 0 = 2.06$ s (3 s.f.) ***[1 mark]***

b) No acceleration horizontally, so the horizontal component of velocity remains constant at $14\cos 46°$ ms^{-1}. ***[1 mark]***

Consider vertical motion once more, taking up as +ve to find the vertical component of the stone's final velocity:

$s = -22;\ u = 14\sin 46°;\ a = -9.8;\ v = ?$

$v^2 = u^2 + 2as = (14\sin 46°)^2 + (2 \times -9.8 \times -22)$ ***[1 mark]***

$= 532.6...$ ***[1 mark]***

Don't bother finding the square root, as you'd only have to square it again in the next bit of the answer.

Speed $= \sqrt{(14\cos 46)^2 + 532.6...}$ ***[1 mark]***

$= 25.0$ ms^{-1} (3 s.f.) ***[1 mark]***

3 a) Consider motion vertically, taking up as positive:

$s = ?,\ u = 10\sin 20°,\ v = 0,\ a = -9.8.$ ***[1 mark]***

Use $v^2 = u^2 + 2as$:

$0 = (10\sin 20°)^2 - 19.6s$ ***[1 mark]***

$\Rightarrow s = (10\sin 20°)^2 \div 19.6 = 0.59682...$ m ***[1 mark]***

The stone is thrown from 1 m above the ground, so the maximum height reached is $1 + 0.59682... = 1.60$ m (3 s.f.) ***[1 mark]***

b) Again consider vertical motion:

$s = -1,\ u = 10\sin 20°,\ a = -9.8,\ t = ?.$

Use $s = ut + \frac{1}{2}at^2$

$-1 = (10\sin 20°)t - 4.9t^2$ ***[1 mark]***

$\Rightarrow 4.9t^2 - (10\sin 20°)t - 1 = 0.$

Use the quadratic formula to find t:

$t = \frac{10\sin 20° + \sqrt{(-10\sin 20°)^2 + (4 \times 4.9 \times 1)}}{9.8}$ ***[1 mark]***

$= 0.91986... = 0.920$ s (3 s.f.) ***[1 mark]***

You don't need to worry about the other value of t that the formula gives you as it will be negative, and the equation isn't valid for t < 0.

c) E.g. Air resistance acting on the stone (affecting acceleration) has not been included in the model, so it may be inaccurate.

[1 mark for any suitable modelling assumption that may affect the accuracy of the model]

Pages 99-104 Forces and Newton's Laws

1 a) Resolve along x-axis:

$P\sin 40° - 20\cos(105 - 90)° = 0$

so, $P = \frac{20\cos 15°}{\sin 40°} = 30.054... = 30.1$ N (3 s.f.)

[2 marks available in total — 1 mark for correct workings, 1 mark for correct value of P]

b) Resolve along y-axis:

$R = 30.054...\cos 40° + 35 - 20\sin 15°$

$= 52.8$ N (3 s.f.) up the y-axis

[3 marks available in total — 1 mark for correct workings, 1 mark for correct value of R and 1 mark for correct direction]

The horizontal component is zero, so the direction is along the y-axis.

2 a) If the resultant force is **R** then

$\mathbf{R} = \mathbf{A} + \mathbf{B} = (2\mathbf{i} - 11\mathbf{j}) + (7\mathbf{i} + 5\mathbf{j}) = (9\mathbf{i} - 6\mathbf{j})$ N ***[1 mark]***

$|\mathbf{R}| = \sqrt{9^2 + (-6)^2}$ ***[1 mark]*** $= 10.8$ N (3 s.f.) ***[1 mark]***

b) Using $\mathbf{F} = m\mathbf{a}$:

$(9\mathbf{i} - 6\mathbf{j}) = 0.5\mathbf{a} \Rightarrow \mathbf{a} = (18\mathbf{i} - 12\mathbf{j})$ ms^{-2}

[2 marks available — 1 mark for correct working, 1 mark for correct answer]

3 Resolving horizontally: $S\cos 40° = F$ ***[1 mark]***

Resolving vertically: $R = 2g + S\sin 40°$ ***[1 mark]***

The ring is stationary, so $F \leq \mu R$ ***[1 mark]***.

So, $S\cos 40° \leq 0.3(2g + S\sin 40°)$ ***[1 mark]***

$S\cos 40° \leq 0.6g + 0.3S\sin 40°$

$S\cos 40° - 0.3S\sin 40° \leq 0.6g$

$S(\cos 40° - 0.3\sin 40°) \leq 0.6g$

$S \leq 10.3$ N (3 s.f.) ***[1 mark]***

4 a)

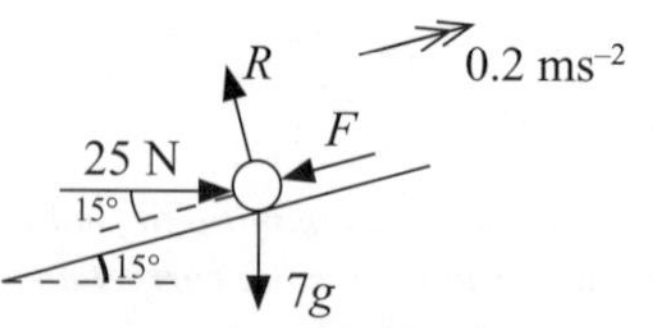

Resolving parallel to the plane (↗):

$F_{net} = ma$

$25\cos 15° - F - 7g\sin 15° = 7 \times 0.2$ ***[1 mark]***

$F = 25\cos 15° - 7g\sin 15° - 1.4$

$F = 4.993...$ N ***[1 mark]***

Resolving perpendicular to the plane (↖):

$F_{net} = ma$

$R - 25\sin 15° - 7g\cos 15° = 7 \times 0$ ***[1 mark]***

$R = 25\sin 15° + 7g\cos 15° = 72.732...$ N

Using $F = \mu R$: ***[1 mark]***

$4.993... = \mu \times 72.732...$ ***[1 mark]***

$\Rightarrow \mu = 0.0686... = 0.07$ (2 d.p.) ***[1 mark]***

b)

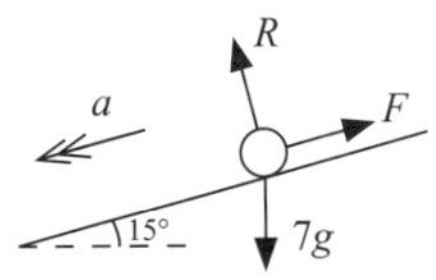

Resolving perpendicular to the plane (↖):
$R = 7g\cos 15°$ ***[1 mark]***
$F = \mu R$ ***[1 mark]***
$F = 0.0686... \times 7g\cos 15°$
Resolving parallel to the plane (↙):
$7g\sin 15° - F = 7a$ ***[1 mark]***
$7g\sin 15° - (0.0686... \times 7g\cos 15°) = 7a$
$a = 1.88...\ \text{ms}^{-2}$ ***[1 mark]***
$s = 3, u = 0, a = 1.88..., t = ?$
$s = ut + \frac{1}{2}at^2$
$3 = \frac{1}{2} \times 1.88... \times t^2$ ***[1 mark]***
$t^2 = 3.18... \Rightarrow t = 1.78$ s (3 s.f.) ***[1 mark]***

5 a)

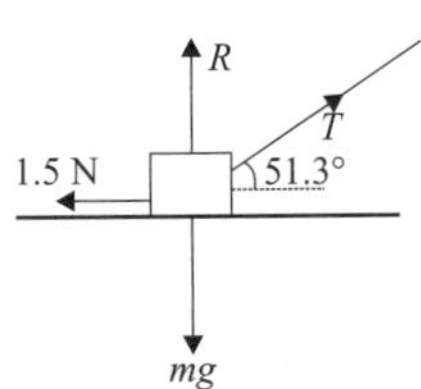

[1 mark]

b) Resolving horizontally:
$F = T\cos 51.3°$ ***[1 mark]***
so $T = \frac{1.5}{\cos 51.3°} = 2.40$ N (3 s.f.) ***[1 mark]***
Resolving vertically:
$mg = R + T\sin 51.3°$ ***[1 mark]***
$R = \frac{F}{\mu} = 1.5 \div 0.6 = 2.5$ N ***[1 mark]***
so $mg = 2.5 + 2.40\sin 51.3°$
and $m = 0.4461...$ kg $= 0.446$ kg (to 3 s.f.) ***[1 mark]***

6

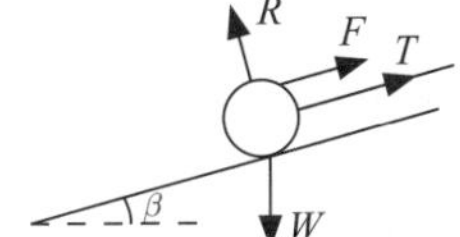

Resolving the forces parallel to the plane (↙):
$W\sin\beta = F + T$ ***[1 mark]***
As the friction is limiting, $F = \mu R$, so
$W\sin\beta = \mu R + T$ ***[1 mark]***
Resolving forces perpendicular to the plane (↖):
$W\cos\beta = R$ ***[1 mark]***
Substitute this in for R in the above equation:
$W\sin\beta = \mu W\cos\beta + T$ ***[1 mark]***
$\Rightarrow \mu W\cos\beta = W\sin\beta - T \Rightarrow \mu = \frac{W\sin\beta}{W\cos\beta} - \frac{T}{W\cos\beta}$
$= \tan\beta - \frac{T}{W}\sec\beta$
[1 mark for using trig identities to rearrange into required form]

7 Call the mass of the woman W.
Drawing two force diagrams will really help:

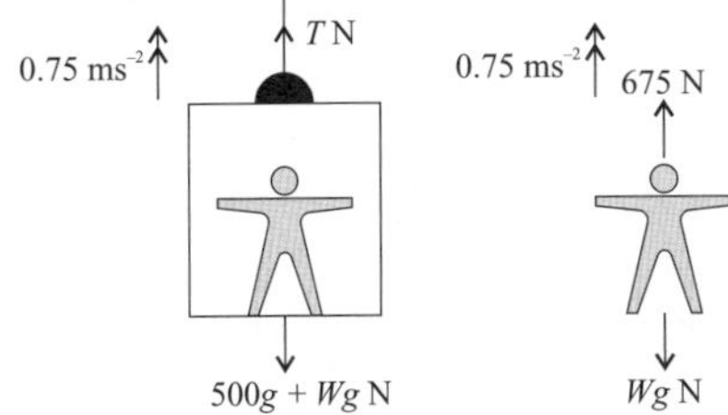

First, use $F = ma$ for the woman in the lift to find her mass:
$675 - Wg = 0.75W \Rightarrow 675 = 0.75W + Wg \Rightarrow W = 63.981...$ kg
Now use $F = ma$ for the whole connected system:
$T - (500 + 63.981...)g = 0.75(500 + 63.981...) \Rightarrow T = 5950$ N
[4 marks available — 1 mark for using F = ma for the woman, 1 mark for finding the mass of the woman, 1 mark for using F = ma on the whole system, 1 mark for finding the correct answer]

8 a) The system is at rest, so equating the forces acting on A gives:
$35g - T = 0 \Rightarrow T = 35g$ ***[1 mark]***
Now do the same for forces acting on B:
$Mg + K - T = 0 \Rightarrow K = T - Mg$
Substitute in the value of T to get:
$K = 35g - Mg = g(35 - M)$ ***[1 mark]***

b) i) $v = 1, u = 0, a = ?$ and $t = 3$. Using $v = u + at$:
$1 = 3a \Rightarrow a = \frac{1}{3}\ \text{ms}^{-2}$ ***[1 mark]***
Using $F = ma$ for A: ***[1 mark]***
$35g - T = 35 \times \frac{1}{3} \Rightarrow T = 35g - \frac{35}{3} = 331.33....$ ***[1 mark]***
Using $F = ma$ for B:
$T - Mg = M \times \frac{1}{3} \Rightarrow T = M(g + \frac{1}{3})$ ***[1 mark]***
$\Rightarrow M = 331.33... \div (g + \frac{1}{3}) = 32.7$ kg (3 s.f.) ***[1 mark]***

ii) E.g. The string is very long, B doesn't reach the pulley, A doesn't reach the ground, acceleration is constant.
[1 mark for a sensible assumption]

9 a)

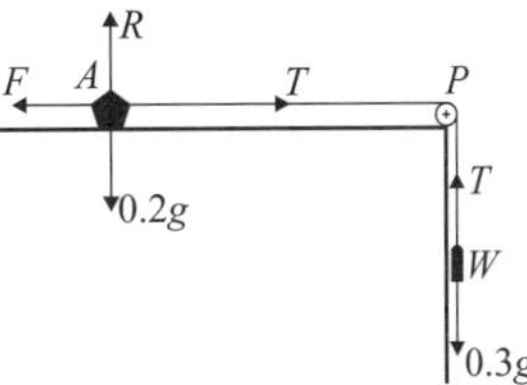

If mass of $A = 0.2$ kg, mass of $W = 1.5 \times 0.2 = 0.3$ kg
For W, $F = ma$:
$0.3g - T = 0.3(4)$ ***[1 mark]***
$T = 2.94 - 1.2 = 1.74$ ***[1 mark]***
For A, $F = ma$:
$T - F = 4(0.2) = 0.8$ ***[1 mark]***
$F = \mu R$, but resolving vertically: $R = 0.2g = 1.96$ ***[1 mark]***
So $F = 1.96\mu$, ***[1 mark]***
so $T - 1.96\mu = 0.8$,
$\Rightarrow 1.74 - 1.96\mu = 0.8$
So $\mu = 0.480$ (3 s.f.) ***[1 mark]***

b) Speed of A at h = speed of W at h (where it impacts ground).
Calculate speed of A at h using $v^2 = u^2 + 2as$:
$v^2 = 0^2 + 2(4 \times h)$ ***[1 mark]***
$v^2 = 8h$ ①
Distance travelled by A beyond $h = \frac{3}{4}h$
Calculate frictional force slowing A after h using $F = ma$:
$F = \mu R$ and $R = mg = 0.2g = 1.96$ ***[1 mark]***
so $F = 0.480 \times 1.96 = 0.94$
so $a = -0.94 \div 0.2 = -4.7\ \text{ms}^{-2}$ ***[1 mark]***
Speed of A at h using $u^2 = v^2 - 2as$:
$u^2 = 3^2 - 2(-4.7 \times \frac{3}{4}h) = 9 + 7.05h$ ***[1 mark]*** ②
Substituting ① into ② (where $v^2 = u^2$):
$8h = 9 + 7.05h$ so $h = 9.47$ m ***[1 mark]***
Time taken to reach h using $s = ut + \frac{1}{2}at^2$:
$9.47 = (0 \times t) + \frac{1}{2}(4 \times t^2) = 2t^2$ ***[1 mark]***
so $t^2 = 4.736...$ and $t = 2.18$ s (3 s.f.) ***[1 mark]***
Tricky. The first thing is to realise that you can calculate the time taken for W to fall indirectly using the connected particle, A. The second thing to realise is that A carries on moving after W hits the ground and is slowed by friction. Then, find that you can calculate the speed of A when it's moved a distance of h in two different ways. Finally, you can find h and use that to find t.

c) E.g. The tension is the same throughout the string — so A and W have the same acceleration when W is falling.
[1 mark for a sensible comment]

10 a) Start by adding values to the diagram:

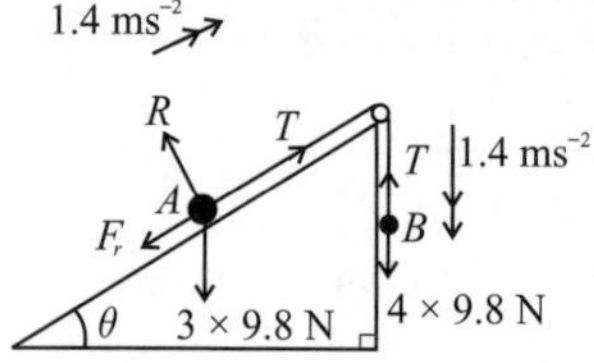

where $\theta = \tan^{-1}\frac{3}{4}$

$\tan\theta = \frac{3}{4} \Rightarrow \cos\theta = \frac{4}{5}$

Resolving forces around A perpendicular to the plane (↖):

$R = 3g\cos\theta = 3 \times 9.8 \times \frac{4}{5} = 23.52$ N ***[1 mark]***

Use $F = ma$ for the vertical motion of B: ***[1 mark]***

$(4 \times 9.8) - T = 4 \times 1.4$

So, $T = 33.6$ N ***[1 mark]***

Use $F = ma$ for A parallel to the plane:

$T - F_r - (3 \times 9.8 \sin\theta) = 3 \times 1.4$ ***[1 mark]***

$\tan\theta = \frac{3}{4} \Rightarrow \sin\theta = \frac{3}{5}$.

Substitute in values of T and $\sin\theta$:

$33.6 - F_r - 17.64 = 4.2$

So, $F_r = 11.76$ N ***[1 mark]***

The system's moving, so friction is 'limiting',

i.e. $F_r = \mu R$ ***[1 mark]***

Rearranging and substituting values in gives:

$\mu = \frac{F_r}{R} = \frac{11.76}{23.52} = 0.5$ ***[1 mark]***

b) Motion of B: $u = 0,\ a = 1.4,\ s = ?,\ t = 2$

Use $s = ut + \frac{1}{2}at^2$: ***[1 mark]***

$s = 0 + \left(\frac{1}{2} \times 1.4 \times 4\right) = 2.8$ m ***[1 mark]***

Particle B moves 2.8 m before the string breaks.

c) While A and B are attached, they move together at the same speed. So you can use the information given to find the speed of (both) A and B when the string breaks:

$v = u + at = 0 + 1.4 \times 2 = 2.8$ ms^{-1} ***[1 mark]***

Draw a diagram to show the forces on A after the string breaks:

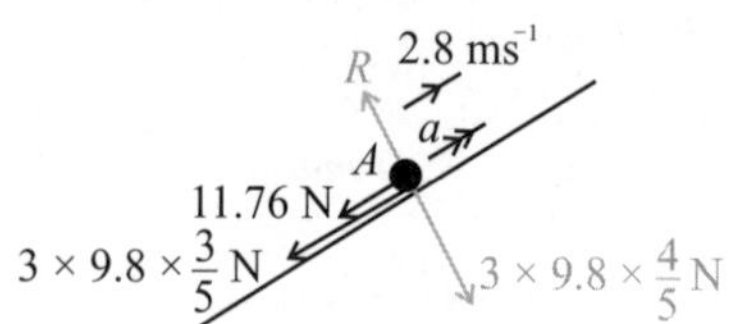

Use $F = ma$ parallel to the plane to find a:

$-11.76 - \left(3 \times 9.8 \times \frac{3}{5}\right) = 3a$ ***[1 mark]***

So, $a = -9.8$ ms^{-2} ***[1 mark]***

Calculate the distance A travels using $v^2 = u^2 + 2as$ ***[1 mark]***

$u = 2.8,\ v = 0,\ a = -9.8,\ s = ?,\ t = ?$

$s = \frac{v^2 - u^2}{2a} = \frac{0 - 2.8^2}{2 \times -9.8} = 0.4$ m ***[1 mark]***

So particle A moves 0.4 m from the instant the string breaks until it comes to rest.

Pages 105-106: Moments

1 Start with a sketch:

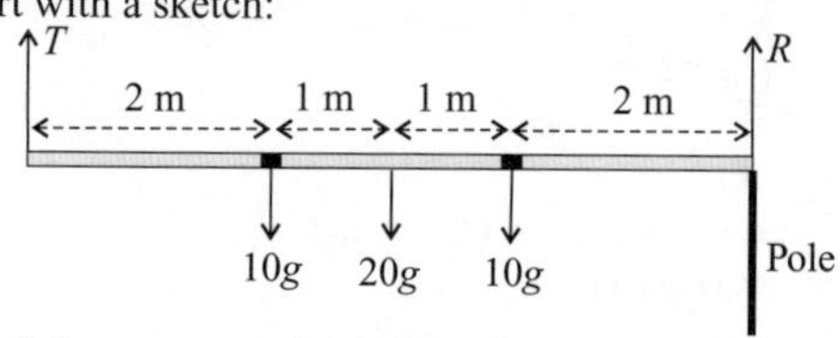

a) Take moments about the pole:

$6T = (2 \times 10g) + (3 \times 20g) + (4 \times 10g)$

$6T = 20g + 60g + 40g = 120g$

So, $T = 20g$

[2 marks available — 1 mark for taking moments about the pole, 1 mark for the correct answer]

You need to take moments around the pole because you don't know what the magnitude of R is — taking moments about the pole means you don't ever need to know.

b) Resolve vertically:

$T + R = 10g + 20g + 10g$

So $20g + R = 40g$

So, $R = 20g$

[2 marks available — 1 mark for resolving vertically, 1 mark for the correct answer]

2 The lamina is uniform, so its centre of mass is at the centre of the rectangle. The perpendicular distance from A to the line of action of the lamina's weight is $3 \div 2 = 1.5$ m ***[1 mark]***.

So, taking moments about A ***[1 mark]***:

$8 \times 1.5 = F \times 3$, so $F = 4$ N ***[1 mark]***

3 a) Start by drawing a diagram:

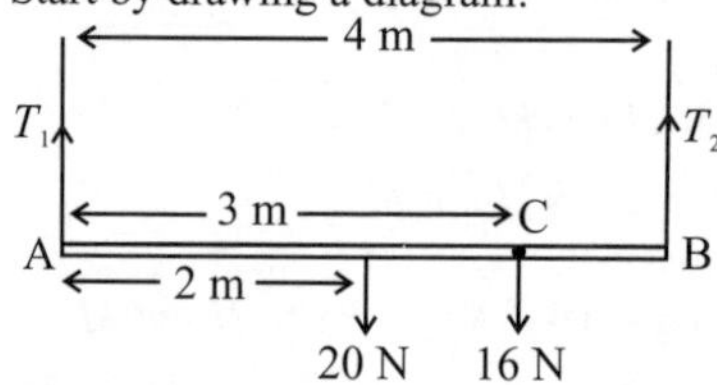

Take moments about A and B, using moments clockwise = moments anticlockwise

A: $(20 \times 2) + (16 \times 3) = T_2 \times 4$

So, $T_2 = 88 \div 4 = 22$

B: $4 \times T_1 = (20 \times 2) + (16 \times 1)$

So, $T_1 = 56 \div 4 = 14$

[3 marks available — 1 mark for taking moments about a point, 1 mark for correct value of A, 1 mark for correct value of B]

You might've resolved forces vertically or taken moments about other points — if you get the same answers, and show full correct working, then you're fine.

b) Start by drawing a new diagram:

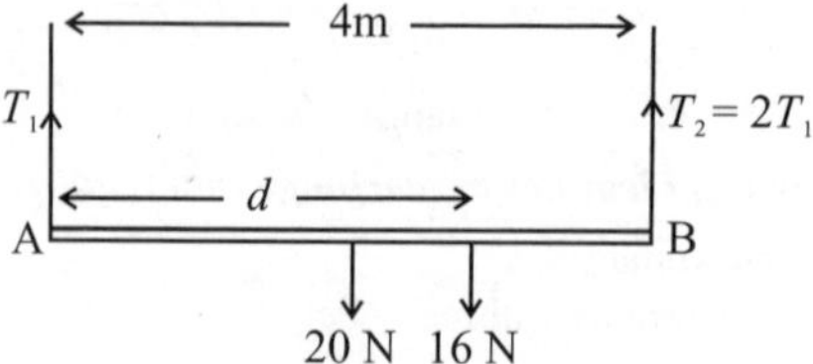

Take moments about A and B:

A: $(20 \times 2) + 16d = 4 \times 2T_1$

So, $T_1 = 5 + 2d$

B: $4T_1 = (20 \times 2) + (4 - d) \times 16$

So, $T_1 = 10 + 16 - 4d = 26 - 4d$

Set the two equations in T_1 equal to each other and solve for d:

$5 + 2d = 26 - 4d \Rightarrow 6d = 21 \Rightarrow d = 3.5$ m

[3 marks available — 1 mark taking moments about a point, 1 mark for attempting to solve simultaneous equations, 1 mark for correct value of d]

4 Let the reaction forces at P and Q be R_P and R_Q respectively. When the beam is on the point of tilting about P or Q, the reaction force at that point will be 0. So, for the beam to be in equilibrium, R_P and R_Q must be ≥ 0 ***[1 mark]***.
If the painter stands between P and Q, then R_P and R_Q must be positive ***[1 mark]***.
(You could show this by taking moments about P and Q but to get the mark, it's enough to just state it as a given.)
Consider the painter standing between L and P, x cm from P.
For the beam to be in equilibrium, clockwise moments about P must be greater than or equal to 0. So take moments about P:
$0.4W - (x \times 5W) \geq 0$ ***[1 mark]***
$\Rightarrow 0.4 \geq 5x \Rightarrow x \leq 0.08$ m $= 8$ cm ***[1 mark]***
Now consider the painter standing between Q and M, y cm from Q.
For the beam to be in equilibrium, anti-clockwise moments about Q must be greater than or equal to 0. So take moments about Q:
$0.7W - (y \times 5W) \geq 0$
$\Rightarrow 0.7 \geq 5y \Rightarrow y \leq 0.14$ m $= 14$ cm ***[1 mark]***
So, the painter can stand anywhere between 32 cm from L to 36 cm from M ***[1 mark for a correct statement]***.

5 a)

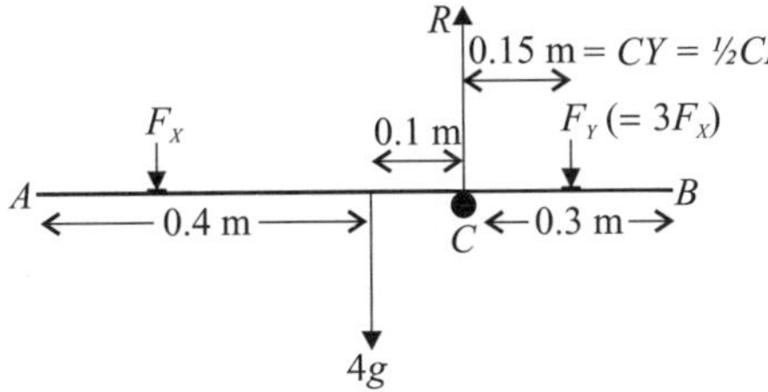

$CY = \frac{1}{2}CB = \frac{1}{2}(0.3) = 0.15$ m
$F_X + F_Y = 80g$, $F_Y = 3 \times F_X$,
so, $4F_X = 80g$, $F_X = 20g$ and $F_Y = 60g$
Taking moments about C:
$F_Y \times CY = (4g \times 0.1) + (F_X \times CX)$
so $(60g \times 0.15) = (4g \times 0.1) + (20g \times CX)$
and $CX = \dfrac{9g - 0.4g}{20g} = 0.43$ m

$AX = 0.5 - CX = 0.07$ m
[5 marks available — 1 mark for finding distances CB and CY, 1 mark for finding correct values of F_X and F_Y, 1 mark for taking moments about a point, 1 mark for correct workings, 1 mark for correct value of AX]

b) $F_X + F_Y = 80g$
Resolving vertically:
$R = F_X + F_Y + 4g = 84g = 823$ N (3 s.f.)
[2 marks available — 1 mark for resolving vertically, 1 mark for correct value of R]

c)

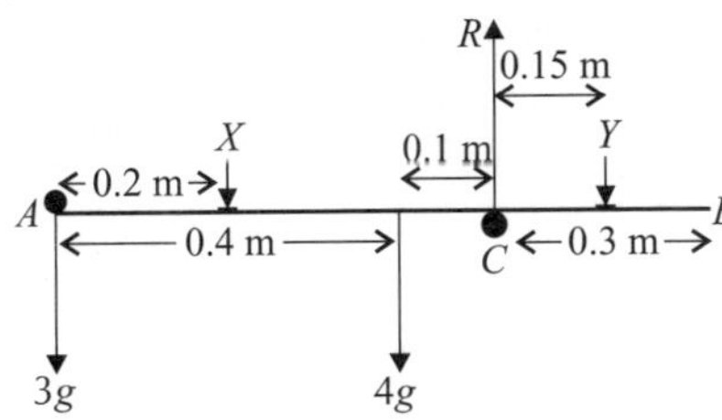

$AX = 0.07 + 0.13 = 0.2$ m
$F_X + F_Y = 80g$
Resolving vertically:
$R = 3g + F_X + 4g + F_Y = 87g$
Taking moments about point X:
$(0.2 \times 3g) + (0.3 \times 87g) = (0.2 \times 4g) + (0.45 \times F_Y)$
so $F_Y = 564$ N
and $F_X = 80g - 564 = 784 - 564 = 220$ N
OR:
Taking moments about point Y:
$(0.65 \times 3g) + (0.45 \times F_X) + (0.25 \times 4g) = (0.15 \times 87g)$
so $F_X = 220$ N
and $F_Y = 80g - 220 = 784 - 220 = 564$ N
[5 marks available — 1 mark for calculating new value of AX, 1 mark for equation connecting F_X and F_Y, 1 mark for taking moments around a point, 1 mark for correct value of F_X, 1 mark for correct value of F_Y]

Practice Exam Paper 1: Pure Mathematics and Mechanics

1 a) If $(x + 1)$ is a factor of $f(x)$, then $f(-1) = 0$:
$f(-1) = (-1)^3 - 2(-1)^2 - 13(-1) - 10 = -1 - 2 + 13 - 10 = 0$
[1 mark], so $(x + 1)$ is factor of $f(x)$ by the factor theorem ***[1 mark]***.

b) $(x + 1)(x^2 - 3x - 10) = 0$ ***[1 mark]***
$(x + 1)(x + 2)(x - 5) = 0$ ***[1 mark]***
$\Rightarrow x = -2, -1$ and 5 ***[1 mark]***

2 Rearrange the equation for the line to get x in terms of y:
$x - 2y + 5 = 0 \Rightarrow x = 2y - 5$ ***[1 mark]***
Substitute the expression for x into the equation for the circle:
$((2y - 5) + 1)^2 + (y - 2)^2 = 5$
$(2y - 4)^2 + (y - 2)^2 = 5$
$4y^2 - 16y + 16 + y^2 - 4y + 4 = 5$
$5y^2 - 20y + 15 = 0$
$y^2 - 4y + 3 = 0$ ***[1 mark]***
$(y - 1)(y - 3) = 0 \Rightarrow y = 1$ or $y = 3$ ***[1 mark for both y-values]***
When $y = 1$, $x = 2(1) - 5 = -3$
When $y = 3$, $x = 2(3) - 5 = 1$ ***[1 mark for both x-values]***
So the points of intersection are $(-3, 1)$ and $(1, 3)$ ***[1 mark]***.
You could have rearranged the first equation to find y in terms of x — but you'd end up with an expression involving fractions, which would make the rest of the working more fiddly.

3 $\tan 3\theta \approx 3\theta$, $\cos 4\theta \approx 1 - \dfrac{16\theta^2}{2} = 1 - 8\theta^2$ and $\sin \theta^2 \approx \theta^2$, so:

$$\frac{1 - \tan 3\theta}{\cos 4\theta - \sin \theta^2} \approx \frac{1 - 3\theta}{1 - 8\theta^2 - \theta^2} = \frac{1 - 3\theta}{1 - 9\theta^2} = \frac{1 - 3\theta}{(1 + 3\theta)(1 - 3\theta)}$$

$$= \frac{1}{1 + 3\theta} \text{ as required}$$

[3 marks available — 1 mark for substituting the correct small angle approximations into the given function, 1 mark for factorising the denominator, 1 mark for correct answer]

4 a)

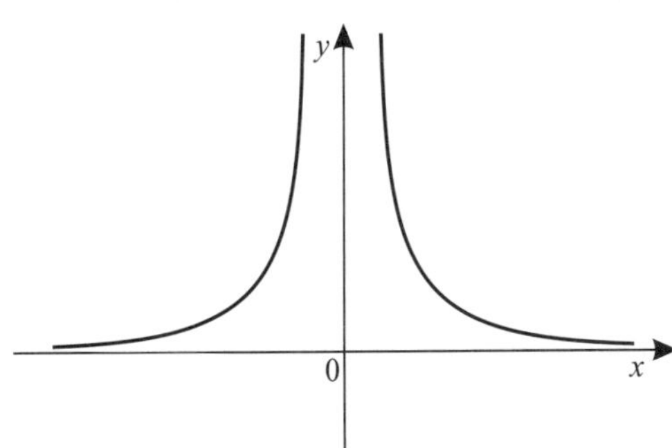

[1 mark for the correct shape]
Asymptotes at $x = 0$ ***[1 mark]*** and $y = 0$ ***[1 mark]***

b)

[1 mark for reflection in the x-axis, 1 mark for translation of +1 in the y-direction)]
Asymptotes at $x = 0$ and $y = 1$ ***[1 mark for both]***

5 $\frac{d}{dx}(5x^3 + 2x^2 + 6) = 15x^2 + 4x$

So $\int \frac{15x^2 + 4x}{5x^3 + 2x^2 + 6}\,dx = \int \frac{f'(x)}{f(x)}\,dx = \ln|f(x)| + C$

$= \ln|5x^3 + 2x^2 + 6| + C$

[3 marks available — 1 mark for recognising that the numerator is the derivative of the denominator, 1 mark for integrating using the correct rule, 1 mark for the correct answer, including the constant]

6 Resolving forces vertically: $R_P + R_Q = 6g + 4g$

$R_P = R_Q$, so: $2R = 98 \Rightarrow R = 49$ N ***[1 mark]***

Taking moments about A: clockwise = anticlockwise

$(6g \times 3) + (4g \times x) = (49 \times 1) + (49 \times 4)$ ***[1 mark]***

$\Rightarrow 176.4 + 39.2x = 245 \Rightarrow x = 1.75$ ***[1 mark]***

You could take moments about any point — as long as your working and answer are correct, you'll get all the marks.

7 a) $\mathbf{v} = \dot{\mathbf{r}} = (3t^2 - 12t + 4)\mathbf{i} + (7 - 8t)\mathbf{j}$

At $t = 5$:

$\mathbf{v} = (3(5^2) - 12(5) + 4)\mathbf{i} + (7 - 8(5))\mathbf{j} = 19\mathbf{i} - 33\mathbf{j}$ as required.

[3 marks available — 1 mark for attempting to differentiate the position vector, 1 mark for a correct expression for velocity at time t, and 1 mark for substituting t = 5 into this expression to obtain the correct velocity]

b) When P is moving south, component of velocity in direction of **i** will be zero, i.e. $3t^2 - 12t + 4 = 0$ ***[1 mark]***

Solve using the quadratic formula:

$t = \frac{-(-12) \pm \sqrt{(-12)^2 - 4 \times 3 \times 4}}{2 \times 3}$

$= 0.36700...$ or $3.63299...$ ***[1 mark]***

When P is moving south, the component of velocity in the direction of **j** will be negative, so find the velocity of P at the two values of t above, and see which is negative:

$\mathbf{v}(0.36700...) = (7 - 8(0.36700...))\mathbf{j} = 4.063...\mathbf{j}$

(so P is moving due north)

$\mathbf{v}(3.63299...) = (7 - 8(3.63299...))\mathbf{j} = -22.06...\mathbf{j}$

So P is moving due south when $t = 3.63$ s (3 s.f.) ***[1 mark]***

Also, when P is moving due south, 7 − 8t must be negative, so $7 - 8t < 0 \Rightarrow t > \frac{7}{8}$, so t cannot be 0.3670...

c) To find force on P, will need to use $\mathbf{F} = m\mathbf{a}$, so first find **a**:

$\mathbf{a} = \dot{\mathbf{v}} = (6t - 12)\mathbf{i} - 8\mathbf{j}$ ***[1 mark]***

When $t = 3$, $\mathbf{a} = (18 - 12)\mathbf{i} - 8\mathbf{j} = 6\mathbf{i} - 8\mathbf{j}$ ***[1 mark]***

So, $\mathbf{F} = m\mathbf{a} = 2.5(6\mathbf{i} - 8\mathbf{j}) = 15\mathbf{i} - 20\mathbf{j}$ ***[1 mark]***

Magnitude of $\mathbf{F} = \sqrt{15^2 + (-20)^2}$ ***[1 mark]***

$= 25$ N ***[1 mark]***

8 a) $y = \frac{2x + 7}{3x - 5} \Rightarrow y(3x - 5) = 2x + 7 \Rightarrow 3xy - 5y = 2x + 7$

$\Rightarrow 3xy - 2x = 7 + 5y \Rightarrow x(3y - 2) = 7 + 5y$

$\Rightarrow x = \frac{7 + 5y}{3y - 2}$

So $f^{-1}(x) = \frac{7 + 5x}{3x - 2}$

[2 marks available — 1 mark for a correct method, 1 mark for the correct answer]

b) $f(2) = \frac{2(2) + 7}{3(2) - 5} = \frac{4 + 7}{6 - 5} = \frac{11}{1} = 11$

$g(11) = 11^2 - k$, so $120 = 121 - k \Rightarrow k = 1$

[3 marks available — 1 mark for correct method for finding f(2), 1 mark for correct value of f(2), 1 mark for the correct value of k]

c) $gh(x) = (\cos x)^2 - 1 = \cos^2 x - 1$ ***[1 mark]***

The range of $\cos^2 x$ is $0 \le \cos^2 x \le 1$,

so the range of $gh(x)$ is $-1 \le gh(x) \le 0$ ***[1 mark]***

$hg(x) = \cos(x^2 - 1)$ ***[1 mark]***

So the range of $hg(x)$ is the range of cos,

i.e. $-1 \le hg(x) \le 1$ ***[1 mark]***

9 a) Set the equations equal to each other and rearrange:

$\frac{1}{x^3} = x + 3 \Rightarrow 1 = x^4 + 3x^3$

$\Rightarrow x^4 + 3x^3 - 1 = 0$ ***[1 mark]***

b) Let $f(x) = x^4 + 3x^3 - 1$, then $f'(x) = 4x^3 + 9x^2$ ***[1 mark]***

Using the Newton-Raphson formula

$x_{n+1} = x_n - \frac{f(x_n)}{f'(x_n)} = x_n - \frac{x_n^4 + 3x_n^3 - 1}{4x_n^3 + 9x_n^2}$ ***[1 mark]***

$x_1 = 0.6$

$x_2 = 0.6 - \frac{0.6^4 + 3(0.6^3) - 1}{4(0.6^3) + 9(0.6^2)}$

$= 0.65419... = 0.6542$ (4 d.p.)

$x_3 = 0.65419... - \frac{0.65419...^4 + 3(0.65419...^3) - 1}{4(0.65419...^3) + 9(0.65419...^2)}$

$= 0.64955... = 0.6496$ (4 d.p.) ***[1 mark for both x_2 and x_3]***

10 a) $\mathbf{u} = (40\mathbf{i} + 108\mathbf{j})$ kmh^{-1}, $\mathbf{a} = (4\mathbf{i} + 12\mathbf{j})$ kmh^{-2}, $t = \frac{1}{2}$ h

Use $\mathbf{s} = \mathbf{u}t + \frac{1}{2}\mathbf{a}t^2$:

$\mathbf{s} = \frac{1}{2}(40\mathbf{i} + 108\mathbf{j}) + \left[\frac{1}{2} \times \left(\frac{1}{2}\right)^2 \times (4\mathbf{i} + 12\mathbf{j})\right]$ ***[1 mark]***

$\mathbf{s} = (20.5\mathbf{i} + 55.5\mathbf{j})$ km ***[1 mark for 20.5i, 1 mark for 55.5j]***

Make sure you keep track of the units you're using — you have to convert the time from minutes to hours to match the units of the velocity and acceleration.

b) $\mathbf{u} = (60\mathbf{i} + 120\mathbf{j})$, $t = \frac{1}{4}$,

$\mathbf{s} = (20.5\mathbf{i} + 55.5\mathbf{j}) - (2.5\mathbf{i} + 25.5\mathbf{j}) = (18\mathbf{i} + 30\mathbf{j})$ ***[1 mark]***

Use $\mathbf{s} = \mathbf{u}t + \frac{1}{2}\mathbf{a}t^2$:

$(18\mathbf{i} + 30\mathbf{j}) = \frac{1}{4}(60\mathbf{i} + 120\mathbf{j}) + \frac{1}{2} \times \left(\frac{1}{4}\right)^2 \times \mathbf{a}$ ***[1 mark]***

$\Rightarrow \frac{1}{32}\mathbf{a} = (18\mathbf{i} + 30\mathbf{j}) - (15\mathbf{i} + 30\mathbf{j})$

$\Rightarrow \mathbf{a} = 32(3\mathbf{i} + 0\mathbf{j}) = 96\mathbf{i}$

So the magnitude of the acceleration is 96 kmh^{-2} ***[1 mark]***.

c) $\mathbf{u} = (30\mathbf{i} - 40\mathbf{j})$, $\mathbf{a} = (5\mathbf{i} + 8\mathbf{j})$, $\mathbf{v} = \mathbf{v}$, $t = ?$, $\mathbf{s} = ?$,

Use $\mathbf{v} = \mathbf{u} + \mathbf{a}t$ to find the time that G is moving parallel to **i** (the unit vector in the direction of east):

$\mathbf{v} = (30\mathbf{i} - 40\mathbf{j}) + t(5\mathbf{i} + 8\mathbf{j})$ ***[1 mark]***

When G is moving parallel to **i**, the **j**-component of its velocity is zero, so: $0 = -40 + 8t$ ***[1 mark]*** $\Rightarrow t = 5$

Check that the i-component is positive: $30 + 5(5) = 55$, so G is moving due east at $t = 5$ ***[1 mark]***.

Now use $\mathbf{s} = \frac{1}{2}(\mathbf{u} + \mathbf{v})t$ to find the position of G at this time:

$\mathbf{s} = \frac{1}{2}(30\mathbf{i} - 40\mathbf{j} + (30\mathbf{i} - 40\mathbf{j}) + 5(5\mathbf{i} + 8\mathbf{j})) \times 5$ ***[1 mark]***

$\mathbf{s} = \frac{5}{2}(85\mathbf{i} - 40\mathbf{j}) = (212.5\mathbf{i} - 100\mathbf{j})$ km ***[1 mark for 212.5i, 1 mark for −100j]***.

11 First find the coordinates of P:

$t = 1 \Rightarrow x = 1^3 + 2 = 3$ and $y = 1^2 + 2 = 3$,

so the coordinates of P are (3, 3) ***[1 mark]***.

$\frac{dx}{dt} = 3t^2$, $\frac{dy}{dt} = 2t$ ***[1 mark for both]***

Using the chain rule: $\frac{dy}{dx} = \frac{dy}{dt} \div \frac{dx}{dt} = \frac{2t}{3t^2} = \frac{2}{3t}$ ***[1 mark]***.

At P, $t = 1$, so $\frac{dy}{dx} = \frac{2}{3}$ ***[1 mark]***.

Using the equation of a straight line formula at P (3, 3):

$y - 3 = \frac{2}{3}(x - 3) \Rightarrow y = \frac{2}{3}x + 1$ ***[1 mark]***.

12 a) Use the product rule: $u = x \Rightarrow \frac{du}{dx} = 1$

$v = \ln x \Rightarrow \frac{dv}{dx} = \frac{1}{x}$ ***[1 mark for both correct]***

$\frac{dy}{dx} = x\left(\frac{1}{x}\right) + \ln x = 1 + \ln x$ ***[1 mark]***

At the stationary point, $\frac{dy}{dx} = 0$

$1 + \ln x = 0$ ***[1 mark]*** $\Rightarrow \ln x = -1 \Rightarrow x = e^{-1}$ ***[1 mark]***

$y = e^{-1} \times \ln e^{-1} = -e^{-1}$

So the stationary point is $(e^{-1}, -e^{-1})$ ***[1 mark]***

b) $\frac{d^2y}{dx^2} = \frac{1}{x}$ ***[1 mark]***

At $(e^{-1}, -e^{-1})$, $\frac{d^2y}{dx^2} = e > 0$, so it is a minimum point ***[1 mark]***

13 a) If the rate of change is proportional to the population, then $\frac{dP}{dt} \propto P \Rightarrow \frac{dP}{dt} = kP$ for some constant k.
Separate the variables and integrate:
$\int \frac{1}{P}\,dP = \int k\,dt \Rightarrow \ln P = kt + \ln Q \Rightarrow P = e^{kt + \ln Q}$
$\Rightarrow P = e^{kt}e^{\ln Q} \Rightarrow P = Qe^{kt}$

[3 marks available — 1 mark for setting up a proportionality statement and converting it into an equation, 1 mark for separating variables and integrating both sides, 1 mark for rearranging to give function in the required form]

b) $Q = 5300$ ***[1 mark]***
$876 = 5300e^{6k}$ ***[1 mark]***
$\Rightarrow \frac{876}{5300} = e^{6k} \Rightarrow \ln\left|\frac{876}{5300}\right| = 6k$
$\Rightarrow 6k = -1.8000... \Rightarrow k = -0.3$ (1 d.p.) ***[1 mark]***

c) The value of k is negative so the population is shrinking ***[1 mark]***.

14 a) Consider vertical motion, taking up as positive:
$a = -g$, $u = U\sin\alpha$ ***[1 mark]***, $v = 0$, $s = ?$:
Use $v^2 = u^2 + 2as$:
$0 = U^2\sin^2\alpha - 2gs$ ***[1 mark]*** $\Rightarrow s = \frac{U^2\sin^2\alpha}{2g}$ ***[1 mark]***.
The golf ball is initially 0.5 m above the ground, so the maximum height, h, it reaches is:
$h = \frac{1}{2} + \frac{U^2\sin^2\alpha}{2g} = \frac{g + U^2\sin^2\alpha}{2g}$ m, as required ***[1 mark]***.

b) You first need to find the horizontal and vertical components of the golf ball's motion when it lands.
Vertically, taking up as positive:
$u = U\sin\alpha$, $a = -g$, $s = -0.5$, $v = v_V$.
Use $v^2 = u^2 + 2as$:
$v_V^2 = U^2\sin^2\alpha + g$ ***[1 mark]***.
Horizontally, $v_H = u_H = U\cos\alpha$, as acceleration is zero ***[1 mark]***.
So, $V = \sqrt{v_V^2 + v_H^2} = \sqrt{U^2\sin^2\alpha + g + U^2\cos^2\alpha}$ ***[1 mark]***
$= \sqrt{U^2(\sin^2\alpha + \cos^2\alpha) + g}$ ***[1 mark]***
$= \sqrt{U^2 + g}$ ms^{-1}, as required ***[1 mark]***.

c) E.g. The effects of air resistance could be factored in. / The beach ball should not be modelled as a particle — its diameter should be taken into account when finding distances.
[1 mark for a sensible suggestion]

15 a) $\theta = \tan^{-1}\left(\frac{3}{4}\right)$ so $\tan\theta = \frac{3}{4}$. This means that the opposite and adjacent sides of a right-angled triangle are 3 and 4, so the hypotenuse is $\sqrt{3^2 + 4^2} = 5$.
This gives $\sin\theta = \frac{3}{5} = 0.6$ and $\cos\theta = \frac{4}{5} = 0.8$
Resolving forces parallel to the plane (↗) for A, where T is the tension in the string:
$F_{net} = ma \Rightarrow T - mg\sin\theta = ma$
$\Rightarrow ma = T - 5.88m$ ① ***[1 mark]***
Resolving forces perpendicular to the plane (↗) for B:
$R - 2mg\cos 70° = 0 \Rightarrow R = 19.6m\cos 70°$ ***[1 mark]***
Friction is limiting, so:
$F = \mu R = 0.45 \times 19.6m\cos 70° = 8.82m\cos 70°$ ***[1 mark]***
Resolving forces parallel to the plane (↘) for B:
$F_{net} = ma \Rightarrow 2mg\sin 70° - T - F = 2ma$ ② ***[1 mark]***
Now consider ① + ②:
$3ma = T - 5.88m + 2mg\sin 70° - T - F$
$= 2mg\sin 70° - 5.88m - 8.82m\cos 70°$ ***[1 mark]***
Divide through by $3m$ to find a:
$a = \frac{2g\sin 70° - 5.88 - 8.82\cos 70°}{3}$
$= 3.1737... = 3.17$ ms^{-2} (3 s.f.) ***[1 mark]***

b) A is on the point of sliding down the plane, so both particles are in equilibrium.
Resolving forces parallel to the plane (↙) for A:
$10g\sin\theta - T = 0 \Rightarrow T = 58.8$ N ***[1 mark]***
Resolving forces perpendicular to the plane (↗) for B:
$R - 2mg\cos 70° = 0 \Rightarrow R = 2mg\cos 70°$
Again, friction is limiting, so:
$F = \mu R = 0.45 \times 2mg\cos 70°$ ***[1 mark]***
Resolving forces parallel to the plane (↖) for B (where the friction now acts down the plane):
$T - F - 2mg\sin 70° = 0$
$\Rightarrow 58.8 - 0.45 \times 2mg\cos 70° - 2mg\sin 70° = 0$ ***[1 mark]***
$\Rightarrow m = \frac{58.8}{2(0.45g\cos 70° + g\sin 70°)} = 2.74$ kg (3 s.f.)***[1 mark]***

Practice Exam Paper 2: Pure Mathematics and Statistics

1 $f(x) = x^2 - 10x + 14 = (x - 5)^2 - 11$
[1 mark for attempting to write in completed square form $(x + p)^2 + q$, 1 mark for the correct values of p and q]
The curve $(x + p)^2 + q$ has a line of symmetry at $x = -p$, so the line of symmetry has equation $x = 5$ ***[1 mark]***.

2 $2\log a^3b - \log ab = \log (a^3b)^2 - \log ab$ ***[1 mark]***
$= \log a^6b^2 - \log ab$
$= \log\frac{a^6b^2}{ab}$ ***[1 mark]***
$= \log a^5b$ ***[1 mark]***

3 Let X represent the number of days in her sample where Chloe has to queue for longer than 2 minutes. Then $X \sim B(50, 0.4)$ ***[1 mark]***
$P(X = 20) = \binom{50}{20}(0.4^{20})(0.6^{30}) = 0.115$ (3 s.f.) ***[1 mark]***
You could also use the binomial functions on your calculator.

4 $|4x + 3| < 11 \Rightarrow -11 < 4x + 3 < 11$ ***[1 mark]***
$-14 < 4x < 8$ ***[1 mark]***
$-3.5 < x < 2$ ***[1 mark]***
You could have solved this one by squaring both sides of the inequality to get $(4x + 3)^2 > 121$ then solving the quadratic inequality.

5 Start with the LHS:
$\sec 2\theta \equiv \frac{1}{\cos 2\theta}$
Use the double angle formula $\cos 2\theta \equiv \cos^2\theta - \sin^2\theta$:
$\equiv \frac{1}{\cos^2\theta - \sin^2\theta}$ ***[1 mark]***
Divide the numerator and denominator by $\cos^2\theta$:
$\equiv \frac{\sec^2\theta}{1 - \tan^2\theta}$ ***[1 mark]***
Use the trig identity $\tan^2\theta \equiv \sec^2\theta - 1$:
$\equiv \frac{\sec^2\theta}{1 - (\sec^2\theta - 1)}$
$\equiv \frac{\sec^2\theta}{2 - \sec^2\theta}$ as required ***[1 mark for correct rearrangement]***

6 Use the quotient rule: $u = \sin x \Rightarrow \frac{du}{dx} = \cos x$ ***[1 mark]***
$v = (2x + 1)^4 \Rightarrow \frac{dv}{dx} = 4(2)(2x + 1)^3 = 8(2x + 1)^3$ ***[1 mark]***
$f'(x) = \frac{(2x + 1)^4\cos x - 8\sin x(2x + 1)^3}{(2x + 1)^8}$
$= \frac{(2x + 1)\cos x - 8\sin x}{(2x + 1)^5}$
[1 mark for use of the quotient rule, 1 mark for the correct answer]

7 $u = \sqrt{x - 1} \Rightarrow \frac{du}{dx} = \frac{1}{2\sqrt{x - 1}}$
$\Rightarrow 2\sqrt{x - 1}\,du = dx \Rightarrow dx = 2u\,du$ ***[1 mark]***
Change the limits: $x = 10 \Rightarrow u = \sqrt{10 - 1} = \sqrt{9} = 3$
$x = 5 \Rightarrow u = \sqrt{5 - 1} = \sqrt{4} = 2$ ***[1 mark for both limits]***
$u^2 = x - 1 \Rightarrow x = u^2 + 1$
$\int_5^{10} \frac{2x}{\sqrt{x - 1}}\,dx = \int_2^3 \frac{2(u^2 + 1)}{u} \times 2u\,du$ ***[1 mark]*** $= \int_2^3 (4u^2 + 4)\,du$
$= \left[\frac{4u^3}{3} + 4u\right]_2^3$ ***[1 mark]***
$= \left(\frac{4}{3}(3^3) + 4(3)\right) - \left(\frac{4}{3}2^3 + 4(2)\right)$
$= 48 - \frac{56}{3} = \frac{88}{3}$ ***[1 mark]***

8 a) Total number of houses = 1524 + 4279 + 851 = 6654
Work out the number of households for each house value group.
Below £150 000: 1524 ÷ 6654 × 100 ≈ 23
Between £150 000 and £250 000: 4279 ÷ 6654 × 100 ≈ 64
Above £250 000: 851 ÷ 6654 × 100 ≈ 13
The researcher would take a random sample of 23 households from the houses valued below £150 000, 64 households from the houses valued between £150 000 and £250 000 and 13 households from the houses valued above £250 000.
[3 marks available — 1 mark for correct method to find the number of households in at least one group, 1 mark for correct number of households in all three groups, 1 mark for a suitable explanation of how to take the stratified sample]

b) The data is grouped into classes, so use the class midpoints to estimate the sample mean and sample standard deviation.
Midpoints of x: 15, 40, 62.5, 87.5, 120 and 170.

sample mean $= \frac{\sum fx}{\sum f} = \frac{8472.5}{100} = 84.725 \approx$ £84.73 ***[1 mark]***

sample standard deviation $= \sqrt{\frac{1}{n-1}\left(\sum fx^2 - n\bar{x}^2\right)}$

$= \sqrt{\frac{1}{99}(834481.25 - 100(84.725)^2)}$ ***[1 mark]***
= 34.325... ≈ £34.33 ***[1 mark]***

c) E.g. Town A has a higher sample mean, which suggests that more money is spent on food by households in town A. ***[1 mark]***
Town B has a smaller sample standard deviation, which suggests the amount of money spent on food in town B is more consistent. ***[1 mark]***

9 a) Using the conditional probability formula:
$P(R|S) = \frac{P(R \cap S)}{P(S)} = \frac{4}{9} \Rightarrow P(R \cap S) = \frac{4}{9}P(S)$ ***[1 mark]***
$P(S|R) = \frac{P(R \cap S)}{P(R)} = \frac{4}{11} \Rightarrow P(R \cap S) = \frac{4}{11}P(R)$ ***[1 mark]***
So $\frac{4}{9}P(S) = \frac{4}{11}P(R) \Rightarrow P(S) = \frac{9}{11}P(R)$ ***[1 mark]***
Now, using the addition law:
$P(R \cup S) = P(R) + P(S) - P(R \cap S)$
$\frac{8}{10} = P(R) + \frac{9}{11}P(R) - \frac{4}{11}P(R)$ ***[1 mark]***
$\frac{8}{10} = \frac{16}{11}P(R) \Rightarrow P(R) = \frac{11}{20} = 0.55$
And $P(S) = \frac{9}{11}P(R) = \frac{9}{11} \times \frac{11}{20} = \frac{9}{20} = 0.45$
[1 mark for both P(R) and P(S)]

b) $P(R \cap S) = \frac{4}{9}P(S) = \frac{4}{9} \times \frac{9}{20} = \frac{1}{5} = 0.2$
$P(R) \times P(S) = \frac{11}{20} \times \frac{9}{20} = \frac{99}{400} = 0.2475$
[1 mark for both probabilities]
So $P(R \cap S) \neq P(R) \times P(S)$, meaning that the events R and S are not independent ***[1 mark]***.
There are other ways to check whether or not these are independent events — for example, you could show that $P(R) \neq P(R \mid S)$.

c) $P(S \cap T) = P(S) \times P(T) = 0.45 \times 0.15 = 0.0675$
Now work everything out using the known probabilities:

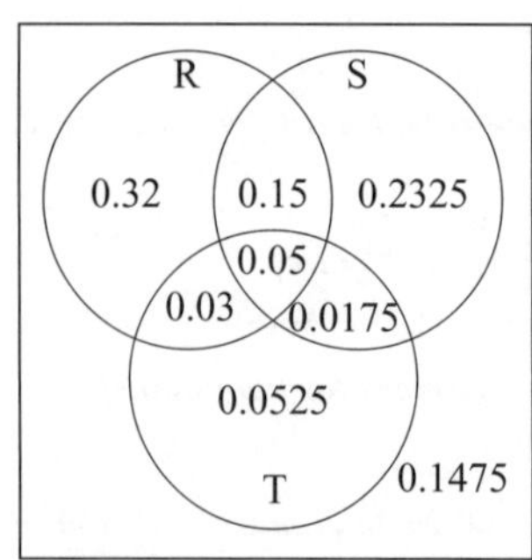

[4 marks for a completely correct Venn diagram, otherwise 1 mark for a correct method to find P(S ∩ T), 1 mark for P(S ∩ T) = 0.0675, 1 mark for P(R ∩ S ∩ T') or P(R ∩ T ∩ S') correct]

10 a) Alice's result shows moderately strong positive correlation, which suggests that the number of bacteria tends to be higher in deeper ponds ***[1 mark]***.
Ben's result is greater than 1, which is not a valid answer, so he must have made an error in his calculation ***[1 mark]***.

b) While the results suggest that they are linked, they do not mean that deeper ponds cause higher numbers of bacteria. There could be a third variable linking them, or it could just be coincidence. ***[1 mark]***

c) $H_0: \rho = 0$, $H_1: \rho > 0$ ***[1 mark]***, so this is a one-tailed test.
The test statistic is $r = 0.71$. The critical value is 0.6215.
Since 0.71 > 0.6215, the result is significant.
There is evidence at the 5% significance level to reject H_0 and to support the alternative hypothesis that the depth of a pond and the number of this type of bacteria in the pond are positively correlated. ***[1 mark]***

11 a) $P(J < 493) = 0.05$ and $P(J > 502) = 0.025$ ***[1 mark]***
$P(J < 493) = 0.05 \Rightarrow P\left(Z < \frac{493 - \mu}{\sigma}\right) = 0.05$
$\Rightarrow \frac{493 - \mu}{\sigma} = -1.645 \Rightarrow \mu - 493 = 1.645\sigma$ ***[1 mark]***
$P(J > 502) = 0.025 \Rightarrow P(J < 502) = 0.975$
$\Rightarrow P\left(Z < \frac{502 - \mu}{\sigma}\right) = 0.975$
$\Rightarrow \frac{502 - \mu}{\sigma} = 1.960 \Rightarrow 502 - \mu = 1.960\sigma$ ***[1 mark]***
Solve these simultaneously:
$(\mu - 493) + (502 - \mu) = 1.645\sigma + 1.960\sigma$
$\Rightarrow 9 = 3.605\sigma \Rightarrow \sigma = 2.496... = 2.50$ (3 s.f.) ***[1 mark]***
$\mu = 493 + 1.645 \times 2.496... = 497.1... = 497$ (3 s.f.) ***[1 mark]***
1.645 and 1.960 came from the percentage points table.

b) From part a), $J \sim N(497, 2.50^2)$ ***[1 mark]***.
The probability that a carton fails to meet the manufacturer's standards is $P(J < 492)$. From your calculator, this is 0.02275... = 0.0228 (4 d.p.) ***[1 mark]***

c) $H_0: \mu = 497$, $H_1: \mu \neq 497$ ***[1 mark for both]***
— so this is a 2-tail test.
Under H_0, $J \sim N(497, 2.50^2)$, so the sample mean,
$\bar{J} \sim N(497, \frac{2.50^2}{20})$ ***[1 mark]***, and $Z = \frac{\bar{J} - 497}{\frac{2.50}{\sqrt{20}}} \sim N(0, 1)$.

$\bar{j} = 498.7$, so $z = \frac{498.7 - 497}{\frac{2.50}{\sqrt{20}}} = 3.0410...$ ***[1 mark]***

This is a 2-tail test, and as $z > 0$, you're interested in the upper tail — so you need to find z such that $P(Z > z) = 0.5\%$.
From the percentage points table, $P(Z > 2.575) = 0.5\%$, so the critical value is 2.575 and the critical region is $Z > 2.575$. Since 3.0410... > 2.575 ***[1 mark]***, the result is significant ***[1 mark]***, so there is evidence at the 1% level of significance to reject H_0 in favour of the alternative hypothesis that the mean volume of juice in a carton has changed ***[1 mark]***.
You could have found P(Z > 3.0410...) instead — you get a value of 0.00117..., which is less than 0.005 so is significant. You'd get the marks for either method — use whichever one you prefer.

12 a) Let M be the midpoint of AB. Then angle $DMB = \frac{\pi}{2}$ radians as $BCDM$ is a rectangle.
Using trigonometry, $DM = 5\sin x$ and $AM = 5\cos x$
$\Rightarrow BC = DM = 5\sin x$ and $AM = MB = CD = 5\cos x$
Perimeter of trapezium, $P = AB + BC + CD + DA$
$= (2 \times 5\cos x) + 5\sin x + 5\cos x + 5$
$= 15\cos x + 5\sin x + 5$
Let $15\cos x + 5\sin x = R\sin(x + \alpha)$
$= R(\sin x\cos\alpha + \cos x\sin\alpha)$
So $R\cos\alpha = 5$ and $R\sin\alpha = 15$
$\Rightarrow \tan\alpha = \frac{15}{5} = 3 \Rightarrow \alpha = \tan^{-1}3 = 1.249...$
$R = \sqrt{15^2 + 5^2} = \sqrt{250} = 5\sqrt{10}$
So $15\cos x + 5\sin x = 5\sqrt{10}\sin(x + 1.249...)$
$\Rightarrow P = 5\sqrt{10}\sin(x + 1.249...) + 5$
So $R = 5\sqrt{10}$ and $\alpha = 1.249...$
[7 marks available — 1 mark for finding an expression for DM, 1 mark for finding an expression for AM, 1 mark for perimeter of trapezium in terms of x, 1 mark for expanding R sin (x + α), 1 mark for the correct value of α, 1 mark for finding the value of R, 1 mark for a correct substitution into the expression for P]

b) The perimeter is 17, so $17 = 5\sqrt{10}\sin(x + 1.249...) + 5$
$\Rightarrow 5\sqrt{10}\sin(x + 1.249...) = 12$
$\Rightarrow \sin(x + 1.249...) = \frac{12}{5\sqrt{10}}$ ***[1 mark]***
As x is acute, $0 \leq x \leq \frac{\pi}{2}$. So you need to find solutions for $1.249... \leq x + 1.249... \leq 2.819...$
$x + 1.249... = \sin^{-1}\left(\frac{12}{5\sqrt{10}}\right)$
$\Rightarrow x + 1.249... = 0.8616...$ (outside range),
$x + 1.249... = \pi - 0.8616... = 2.279...$ ***[1 mark]***
$\Rightarrow x = 2.279... - 1.249... = 1.030... = 1.03$ radians (3 s.f.)
[1 mark]

13 a) You need to find the total area of the histogram first.
The total of the bars' areas is:
$(5 \times 11.2) + (5 \times 23.2) + (10 \times 14.8) + (25 \times 3.2) = 400$
This represents 200 countries, so each country is represented by an area of 2 ***[1 mark]***.
The area of the bar showing birth rates between 25 and 50
$25 \times 3.2 = 80$. This represents $80 \div 2 = 40$ countries ***[1 mark]***.
30 is one-fifth of the way through the 25-50 class, so approximately $40 \times \frac{4}{5} = 32$ countries have a higher birth rate than Eritrea ***[1 mark]***.

b) (i) E.g. The distribution for Africa is negatively skewed whereas the distribution for Europe is fairly symmetrical.
The birth rates in Europe have a lower median and all values for Europe are significantly lower, which shows that birth rates are generally lower in Europe.
The birth rates in Africa have a higher range and interquartile range, which shows that there is more variation in birth rates across all the countries in Africa.
[1 mark for each sensible comment, up to a total of 3 marks]
(ii) E.g. The distribution for African countries is not symmetrical/ negatively skewed, so a normal distribution would not be appropriate ***[1 mark]***.

c) Outliers are at least 2 times the standard deviation away from the mean.
mean – 2 × s.d. = 19.75 – 2 × 6.22 = 7.31 ***[1 mark]***
mean + 2 × s.d. = 19.75 + 2 × 6.22 = 32.19 ***[1 mark]***
The minimum 9.84 > 7.31 and the maximum 31.45 < 32.19 so there are no outliers for the countries in the Middle East ***[1 mark]***.

d) (i) Negative correlation ***[1 mark]***
(ii) Substitute 24.67 into the equation for the regression line:
GDP per capita = 19.438 – 0.438 × 24.67 ***[1 mark]*** = 8.63254
So the GDP per capita for Swaziland is approximately:
8.63254 × 1000 = £8632.54 ***[1 mark]***
(iii) The estimate in part (ii) is interpolation (since 24.67 is within the range of the data), so it should be reliable ***[1 mark]***.

e) (i) and (ii):

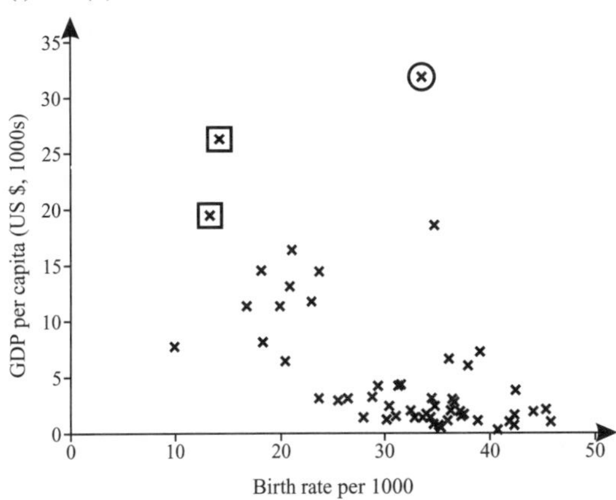

[1 mark for putting a circle around the correct point]
[1 mark for putting squares around both correct points]
Country A has a life expectancy similar to other African countries so you'd expect it to have a similar birth rate, but it has a high GDP per capita because of its oil production. Countries B and C are islands so their birth rates could be different to other African countries, and you'd expect them to have high GDP because they are luxury tourist destinations.

14 Differentiate the equation:
$2x - y - x\frac{dy}{dx} = 6y^2\frac{dy}{dx} \Rightarrow \frac{dy}{dx} = \frac{2x - y}{x + 6y^2}$
To find the gradient, substitute $x = 2$ and $y = 1$ into the expression for $\frac{dy}{dx}$:
$\frac{dy}{dx} = \frac{2(2) - 1}{2 + 6(1)^2} = \frac{3}{8}$
[5 marks available — 1 mark for attempting implicit differentiation, 1 mark for differentiating all terms correctly, 1 mark for the correct expression for dy/dx, 1 mark for substituting x = 2 and y = 1, 1 mark for the correct answer]

15 a) $\frac{dT}{dt}$ is the rate of heat loss with respect to time.
The difference in temperature is $(T - 15)$ °C.
$\frac{dT}{dt} \propto (T - 15)$, so $\frac{dT}{dt} = -k(T - 15)$, with the negative sign indicating that the temperature is decreasing ***[1 mark]***.

b) Rearrange and integrate both sides of the equation from a)
$\int \frac{1}{T - 15}\,dT = \int -k\,dt$
$\ln|T - 15| = -kt + C$ ***[1 mark]***
$T - 15 = e^{-kt + C}$ ***[1 mark]*** $= Ae^{-kt}$ (where $A = e^C$)
Put in the initial condition $t = 0$, $T = 95$ to find A:
$95 - 15 = Ae^0 \Rightarrow A = 80$
$T = 80e^{-kt} + 15$ ***[1 mark]***

c) Put in the given condition $t = 10$, $T = 55$ to find k:
$55 = 80e^{-10k} + 15$ ***[1 mark]*** $\Rightarrow e^{-10k} = 0.5$
$\Rightarrow -10k = \ln 0.5 \Rightarrow k = -\frac{\ln 0.5}{10}$ ***[1 mark]***
$T = 80e^{\frac{\ln 0.5}{10} \times t} + 15$ ***[1 mark]*** $= 80(e^{\ln 0.5})^{\frac{t}{10}} + 15$
$\Rightarrow T = 80(0.5)^{\frac{t}{10}} + 15$ ***[1 mark]***

d) E.g. the model assumes that the temperature in the kitchen remains constant ***[1 mark for any sensible assumption]***.

Practice Exam Paper 3: Pure Mathematics and Comprehension

1 a) $\int_0^{2\pi} 2\cos\frac{x}{4}\,dx$ is the area under the curve between $x = 0$ and $x = 2\pi$. The area of a rectangle with vertices (0, 0) (0, 2) $(2\pi, 0)$ and $(2\pi, 2)$ would have area $2\pi \times 2 = 4\pi$.
The area beneath the curve and between $x = 0$ and $x = 2\pi$ fits inside this rectangle and has an area less than the rectangle, so Lily's answer must be too big ***[1 mark]***.

b) $\int_0^{2\pi} 2\cos\frac{x}{4}\,dx = \left[8\sin\frac{x}{4}\right]_0^{2\pi} = \left[8\sin\frac{\pi}{2}\right] - [8\sin 0] = 8$
[3 marks available — 1 mark for integrating correctly, 1 mark for correct handling of the limits, 1 mark for correct answer]

2 a) The dog's range is a sector, bounded by the edge of the pen, the fence and the arc that is formed by the limit of its rope.
The angle of the sector is the exterior angle of the regular hexagon: $360° \div 60° = \frac{\pi}{3}$ radians ***[1 mark]***
So the area is $\frac{1}{2} \times 2^2 \times \frac{\pi}{3}$ ***[1 mark]*** $= \frac{2\pi}{3}$ m² ***[1 mark]***

b) The hexagon is made up of 6 equilateral triangles of side length 2 m. Area of one of these triangles:
$\frac{1}{2} \times 2 \times 2 \times \sin\left(\frac{\pi}{3}\right)$ ***[1 mark]*** $= 2 \times \frac{\sqrt{3}}{2} = \sqrt{3}$ m²
So the hexagon's area is $6\sqrt{3}$ m² ***[1 mark]***
This means the cat can go in an area of
$(20 \times 8) - \frac{2\pi}{3} - 6\sqrt{3}$ ***[1 mark]*** $= 148$ m² (3 s.f.) ***[1 mark]***

3 a) The number of minutes of practice on the nth day is given by
$u_n = a + (n-1)d = 60 + (n-1)10 = 50 + 10n$ ***[1 mark]***
4 hours 40 minutes is 280 minutes, so
$50 + 10n = 280$ ***[1 mark]*** $\Rightarrow n = 23$, i.e. day 23 ***[1 mark]***

b) The number of minutes of practice on the nth day is given by
$u_n = ar^{n-1} = 100 \times 1.04^{n-1}$ ***[1 mark]***
$100 \times 1.04^{n-1} > 280 \Rightarrow 1.04^{n-1} > 2.8$ ***[1 mark]***
Taking logs gives: $\log 1.04^{n-1} > \log 2.8$
$\Rightarrow (n-1)\log 1.04 > \log 2.8$ ***[1 mark]***
$\Rightarrow n - 1 > \frac{\log 2.8}{\log 1.04} \Rightarrow n > 27.3$ (3 s.f.)
So day 28 ***[1 mark]***

c) Pianist: $\frac{1}{2} \times 30[(2 \times 60) + (29 \times 10)] = 6150$ minutes ***[1 mark]***
Violinist: $\frac{100(1.04^{30} - 1)}{1.04 - 1} = 5608.5$ minutes ***[1 mark]***
The pianist will practise for longer over the 30 days ***[1 mark]***.

4 a) Rearrange the parametric equations:
$x - 1 = 5\cos t \Rightarrow (x-1)^2 = 25\cos^2 t$ ***[1 mark]***
$y - 2 = 5\sin t \Rightarrow (y-2)^2 = 25\sin^2 t$ ***[1 mark]***
$(x-1)^2 + (y-2)^2 = 25\cos^2 t + 25\sin^2 t$
$= 25(\sin^2 t + \cos^2 t) = 25$ ***[1 mark]***

b) Gradient of the radius from the centre (1, 2) to (4, 6):
$\frac{6-2}{4-1} = \frac{4}{3}$ ***[1 mark]***
So the gradient of the tangent at P is $-1 \div \frac{4}{3} = -\frac{3}{4}$ ***[1 mark]***
Using the equation of a straight line:
$y - 6 = -\frac{3}{4}(x - 4)$ ***[1 mark]*** $\Rightarrow 3x + 4y - 36 = 0$ ***[1 mark]***
(so $a = 3$, $b = 4$ and $c = -36$)

5 a) Use the chain rule on the e^{x^2} term:
Let $u = x^2$, then $y = e^u \Rightarrow \frac{du}{dx} = 2x$, $\frac{dy}{du} = e^u$
$\frac{dy}{dx} = \frac{dy}{du} \times \frac{du}{dx} = e^u \times 2x = 2xe^{x^2}$
So for $y = 4x^3 + e^{x^2}$, $\frac{dy}{dx} = 12x^2 + 2xe^{x^2}$
[3 marks available — 1 mark for using the chain rule to differentiate e^{x^2}, 1 mark for each correct term in the answer]

b) Differentiate again to find the second derivative:
$\frac{d}{dx}12x^2 = 24x$
Use the product rule to differentiate $2xe^{x^2}$:
Let $u = 2x$, then $\frac{du}{dx} = 2$
Let $v = e^{x^2}$, then $\frac{dv}{dx} = 2xe^{x^2}$ (from part a)).
$\frac{dy}{dx} = u\frac{dv}{dx} + v\frac{du}{dx} = 2x(2xe^{x^2}) + e^{x^2}(2) = 4x^2e^{x^2} + 2e^{x^2}$
[1 mark for using the product rule correctly]
So $\frac{d^2y}{dx^2} = 24x + 4x^2e^{x^2} + 2e^{x^2}$
[1 mark for correct second derivative]
The curve is concave downwards when $\frac{d^2y}{dx^2} < 0$,
i.e. when $24x + 4x^2e^{x^2} + 2e^{x^2} < 0$ ***[1 mark]***
$e^{x^2} > 0$ and $4x^2 \geq 0$ ***[1 mark]***, so $24x + 4x^2e^{x^2} + 2e^{x^2} < 0$ when $24x < 0$, i.e. when $x < 0$ ***[1 mark]***.

6 a) $\frac{11x - 7}{(2x-4)(x+1)} \equiv \frac{A}{2x-4} + \frac{B}{x+1}$ ***[1 mark]***
$\Rightarrow 11x - 7 \equiv A(x+1) + B(2x-4)$ ***[1 mark]***
Putting in $x = -1$ gives: $-18 = -6B \Rightarrow B = 3$
Putting in $x = 2$ gives: $15 = 3A \Rightarrow A = 5$
So $\frac{11x-7}{(2x-4)(x+1)} \equiv \frac{5}{2x-4} + \frac{3}{x+1}$.
[1 mark for each correct fraction]

b) $\frac{11x-7}{(2x-4)(x+1)} \equiv 5(2x-4)^{-1} + 3(x+1)^{-1}$ ***[1 mark]***
$5(2x-4)^{-1} = 5\left((-4)^{-1}\left(-\frac{x}{2} + 1\right)^{-1}\right)$
$= 5\left(-\frac{1}{4}\left(1 + -1\left(-\frac{x}{2}\right) + \frac{-1\times-2}{1\times2}\left(-\frac{x}{2}\right)^2 + \ldots\right)\right)$ ***[1 mark]***
$= -\frac{5}{4} - \frac{5}{8}x - \frac{5}{16}x^2 + \ldots$
$3(x+1)^{-1} = 3\left(1 + -1(x) + \frac{-1\times-2}{1\times2}x^2 + \ldots\right)$ ***[1 mark]***
$= 3 - 3x + 3x^2 + \ldots$
$\frac{11x-7}{(2x-4)(x+1)} \equiv -\frac{5}{4} - \frac{5}{8}x - \frac{5}{16}x^2 + 3 - 3x + 3x^2 + \ldots$
$\equiv \frac{7}{4} - \frac{29}{8}x + \frac{43}{16}x^2 + \ldots$
[1 mark for each correct term]

7 Use the identity $\cot^2 x + 1 \equiv \text{cosec}^2 x$ and substitute it into the equation:
$2(\cot^2 x + 1) + 5\cot x = 9 \Rightarrow 2\cot^2 x + 5\cot x - 7 = 0$ ***[1 mark]***
Factorising gives $(2\cot x + 7)(\cot x - 1) = 0$ ***[1 mark]***
$\cot x = -\frac{7}{2}$ or $\cot x = 1$
$\Rightarrow \tan x = -\frac{2}{7}$ or $\tan x = 1$ ***[1 mark]***
$x = -15.9°$ (1 d.p.) ***[1 mark]*** and $x = 45°$ ***[1 mark]***.

8 a) Volume $= x^3$, so the price of concrete is $65x^3$ ***[1 mark]***
Surface area $= 6x^2$, so the price of gold leaf is $900x^2$ ***[1 mark]***
Total cost of materials is £250, so $65x^3 + 900x^2 = 250$
[1 mark]
So $900x^2 = 250 - 65x^3 \Rightarrow x^2 = \frac{250 - 65x^3}{900}$
$\Rightarrow x = \sqrt{\frac{250 - 65x^3}{900}}$
[1 mark for correct rearrangement]

b) Let $f(x) = x - \sqrt{\frac{250 - 65x^3}{900}}$
If x, the length of the cube, is between 0.5 and 1, there will be a root of $f(x) = 0$ in the interval $0.5 < x < 1$, and a change of sign for $f(x)$ between 0.5 and 1.
$f(0.5) = 0.5 - \sqrt{\frac{250 - 65(0.5)^3}{900}} = -0.018\ldots$
$f(1) = 1 - \sqrt{\frac{250 - 65(1)^3}{900}} = 0.546\ldots$ ***[1 mark for both]***
There is a change of sign and the function is continuous over this interval, so there is root in the interval $0.5 < x < 1$ ***[1 mark]***.

c) $x_2 = \sqrt{\frac{250 - 65(0.75)^3}{900}} = 0.49730... = 0.4973$ (4 d.p.)
[1 mark]
$x_3 = 0.51855...$
$x_4 = 0.51740...$
$x_5 = 0.51746... = 0.5175$ (4 d.p.) ***[1 mark]***

d) 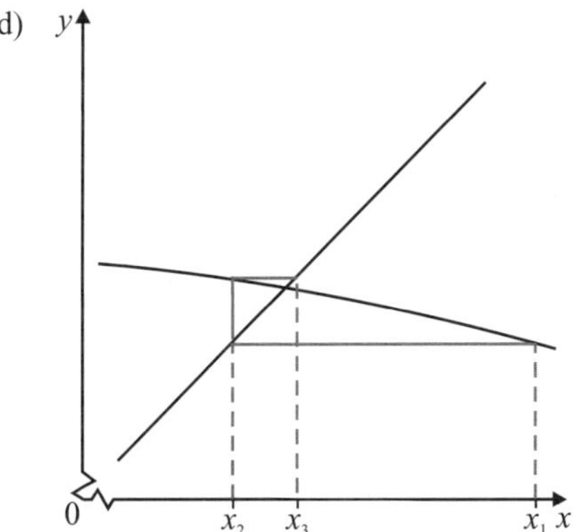

[1 mark for positions of x_2 and x_3 correct]
This is a convergent cobweb diagram.

9 Let a, b, h be a Pythagorean triple, such that $a^2 + b^2 = h^2$, and let n be any positive integer. Multiplying each of a, b and h by n gives na, nb and nh.
$(na)^2 + (nb)^2 = n^2a^2 + n^2b^2 = n^2(a^2 + b^2) = n^2h^2$,
so $(na)^2 + (nb)^2 = (nh)^2$.
Therefore na, nb and nh form a Pythagorean triple.
[2 marks available — 1 mark for multiplying a, b and h by the same integer and squaring, 1 mark for showing this forms a Pythagorean triple]

10 You need to find two consecutive squares with a difference between them of $a^2 = 49$. The differences between consecutive squares form an arithmetic sequence whose n^{th} term is $2n - 1$ (where the nth term is the difference between n^2 and $(n - 1)^2$).
If $2n - 1 = 49$, then $n = 25$, and so $25^2 - 24^2 = 49$ ***[1 mark]***.
This means the Pythagorean triple is 7, 24, 25 ***[1 mark]***.

11 Call the smaller of the two integers p.
Then $\frac{1}{p} + \frac{1}{p+2} = \frac{9}{40}$ ***[1 mark]***.
Multiply both sides by $40p(p + 2)$:
$40(p + 2) + 40p = 9p(p + 2)$ ***[1 mark]***
This gives: $9p^2 - 62p - 80 = 0$, or $(9p + 10)(p - 8) = 0$.
So $p = -\frac{10}{9}$ (which is not an integer, so can be rejected) or $p = 8$.
This means the starting numbers were 8 and 10 ***[1 mark]***.

12 a) If the smaller of the two numbers is $n - 1$ (where n is an integer greater than 1), then the larger must be $n - 1 + 2 = n + 1$.
The sum of the unit fractions is:
$\frac{1}{n-1} + \frac{1}{n+1} = \frac{2n}{(n-1)(n+1)} = \frac{2n}{n^2-1}$ ***[1 mark]***
So $2n$ and $n^2 - 1$ are two legs of the triple.
$(2n)^2 + (n^2 - 1)^2 = 4n^2 + n^4 - 2n^2 + 1 = n^4 + 2n^2 + 1$
$= (n^2 + 1)^2$ ***[1 mark]***.
This is the square of an integer ***[1 mark]***, and so $2n$, $n^2 - 1$ and $n^2 + 1$ form a Pythagorean triple.

b) x is even, so $x = 2n \Rightarrow n = \frac{x}{2}$.
y is the other leg, so $y = n^2 - 1$.
So $y = n^2 - 1 = \left(\frac{x}{2}\right)^2 - 1$ ***[1 mark]***.

13 a) If d is even, then d can be written as $d = 2k$, for some integer k.
So $d^2 = (2k)^2 = 4k^2$, which means that d^2 must be divisible by 4 ***[1 mark]***.

b) Suppose a and b are both odd. Then a can be written $a = 2m + 1$ for some integer m, and so
$a^2 = (2m + 1)^2 = 4m^2 + 4m + 1$.
Similarly, b can be written $b = 2n + 1$ for some integer n, and so $b^2 = (2n + 1)^2 = 4n^2 + 4n + 1$.
Since $a^2 + b^2 = h^2$, this means
$h^2 = 4m^2 + 4m + 1 + 4n^2 + 4n + 1$
$= 2(2m^2 + 2m + 2n^2 + 2n + 1)$, which means h^2 is even.
From the proof given in the comprehension material, this also means that h is even ***[1 mark]***.
However, $h^2 = 4(m^2 + m + n^2 + n) + 2$, which is not divisible by 4 ***[1 mark]***. From part a), this means that h cannot be even.
This is a contradiction, so a and b cannot both be odd ***[1 mark]***.

Formula Sheet

These formulas are the ones you'll be given in your exam.
Make sure you know exactly **when you need them** and **how to use them**.

Arithmetic Series

$$S_n = \frac{1}{2}n(a + l)$$
$$= \frac{1}{2}n[2a + (n - 1)d]$$

Geometric Series

$$S_n = \frac{a(1 - r^n)}{1 - r}$$
$$S_\infty = \frac{a}{1 - r} \text{ for } |r| < 1$$

Binomial Series

$$(a + b)^n = a^n + {}^nC_1 a^{n-1}b + {}^nC_2 a^{n-2}b^2 + \ldots + {}^nC_r a^{n-r}b^r + \ldots + b^n \quad (n \in \mathbb{N})$$

$$\text{where } {}^nC_r = \binom{n}{r} = \frac{n!}{r!(n - r)!}$$

$$(1 + x)^n = 1 + nx + \frac{n(n - 1)}{2!}x^2 + \ldots + \frac{n(n - 1)\ldots(n - r + 1)}{r!}x^r + \ldots \quad (|x| < 1, n \in \mathbb{R})$$

Differentiation

f(x)	**f'(x)**
$\tan kx$	$k \sec^2 kx$
$\sec x$	$\sec x \tan x$
$\cot x$	$-\text{cosec}^2 x$
$\text{cosec } x$	$-\text{cosec } x \cot x$

Quotient rule: for $y = \frac{u}{v}$, $\frac{dy}{dx} = \frac{v\frac{du}{dx} - u\frac{dv}{dx}}{v^2}$

Differentiation from first principles: $f'(x) = \lim_{h \to 0} \frac{f(x + h) - f(x)}{h}$

Integration

$$\int \frac{f'(x)}{f(x)}\,dx = \ln|f(x)| + c$$

$$\int f'(x)(f(x))^n\,dx = \frac{1}{n + 1}(f(x))^{n+1} + c$$

Integration by parts: $\int u\frac{dv}{dx}dx = uv - \int v\frac{du}{dx}dx$

Formula Sheet

Trigonometry

Small angle approximations: $\sin\theta \approx \theta$, $\cos\theta \approx 1 - \frac{1}{2}\theta^2$, $\tan\theta \approx \theta$, where θ is measured in radians.

Trigonometric identities:

$$\sin(A \pm B) \equiv \sin A \cos B \pm \cos A \sin B$$

$$\cos(A \pm B) \equiv \cos A \cos B \mp \sin A \sin B$$

$$\tan(A \pm B) \equiv \frac{\tan A \pm \tan B}{1 \mp \tan A \tan B} \quad (A \pm B \neq (k + \tfrac{1}{2})\pi)$$

Numerical Methods

Trapezium rule: $\int_a^b y\,\mathrm{d}x \approx \frac{1}{2}h\{(y_0 + y_n) + 2(y_1 + y_2 + ... + y_{n-1})\}$, where $h = \frac{b-a}{n}$

The Newton-Raphson iteration for solving $\mathrm{f}(x) = 0$: $x_{n+1} = x_n - \frac{\mathrm{f}(x_n)}{\mathrm{f}'(x_n)}$

Probability

$$\mathrm{P(A \cup B) = P(A) + P(B) - P(A \cap B)}$$

$$\mathrm{P(A \cap B) = P(A)P(B|A) = P(B)P(A|B)} \quad \text{or} \quad \mathrm{P(A|B) = \frac{P(A \cap B)}{P(B)}}$$

Sample Variance

$$s^2 = \frac{1}{n-1}S_{xx}, \text{ where } S_{xx} = \sum(x_i - \overline{x})^2 = \sum x_i^2 - \frac{(\sum x_i)^2}{n} = \sum x_i^2 - n\overline{x}^2$$

Standard deviation, $s = \sqrt{\text{variance}}$

The Binomial Distribution

If $X \sim \mathrm{B}(n, p)$, then $\mathrm{P}(X = r) = {}^n\mathrm{C}_r p^r q^{n-r}$, where $q = 1 - p$

Mean of $X = np$

Formula Sheet

Hypothesis Testing for the Mean of a Normal Distribution

If $X \sim N(\mu, \sigma^2)$, then $\overline{X} \sim N\left(\mu, \frac{\sigma^2}{n}\right)$ and $\frac{\overline{X} - \mu}{\sigma / \sqrt{n}} \sim N(0, 1)$

Percentage Points of the Normal Distribution

p	10	5	2	1
z	1.645	1.960	2.326	2.575

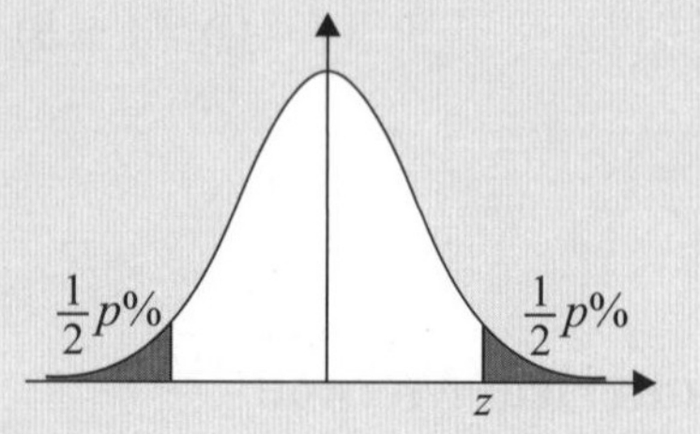

Kinematics

Motion in a straight line:

$$v = u + at$$

$$s = ut + \frac{1}{2}at^2$$

$$s = \frac{1}{2}(u + v)t$$

$$v^2 = u^2 + 2as$$

$$s = vt - \frac{1}{2}at^2$$

Motion in two dimensions:

$$\mathbf{v} = \mathbf{u} + \mathbf{a}t$$

$$\mathbf{s} = \mathbf{u}t + \frac{1}{2}\mathbf{a}t^2$$

$$\mathbf{s} = \frac{1}{2}(\mathbf{u} + \mathbf{v})t$$

$$\mathbf{s} = \mathbf{v}t - \frac{1}{2}\mathbf{a}t^2$$

MRMQ71